STATISTICS DONE WRONG

STATISTICS DONE WRONG

The Woefully Complete Guide

by Alex Reinhart

no starch press

San Francisco

19 18 17 16 15   1 2 3 4 5 6 7 8 9

ISBN-10: 1-59327-620-6
ISBN-13: 978-1-59327-620-1

Publisher: William Pollock
Production Editor: Alison Law
Cover Illustration: Josh Ellingson
Developmental Editors: Greg Poulos and Leslie Shen
Technical Reviewer: Howard Seltman
Copyeditor: Kim Wimpsett
Compositor: Alison Law
Proofreader: Emelie Burnette

For information on distribution, translations, or bulk sales,
please contact No Starch Press, Inc. directly:

No Starch Press, Inc.
245 8th Street, San Francisco, CA 94103
phone: 415.863.9900; info@nostarch.com
www.nostarch.com

Library of Congress Cataloging-in-Publication Data
Reinhart, Alex, 1991-
  Statistics done wrong : the woefully complete guide / by Alex Reinhart.
      pages cm
  Includes index.
  Summary: "Discusses how to avoid the most common statistical errors
  in modern research, and perform more accurate statistical analyses"
  - Provided by publisher.
  ISBN 978-1-59327-620-1 - ISBN 1-59327-620-6
 1. Statistics-Methodology. 2. Missing observations (Statistics) I. Title.
 QA276.R396 2015
 519.5-dc23
                                        2015002128

The first principle is that you must not fool yourself,
and you are the easiest person to fool.

—Richard P. Feynman

To consult the statistician after an experiment is finished is
often merely to ask him to conduct a post mortem examination.
He can perhaps say what the experiment died of.

—R.A. Fisher

About the Author

Alex Reinhart is a statistics instructor and PhD student at Carnegie Mellon University. He received his BS in physics at the University of Texas at Austin and does research on locating radioactive devices using physics and statistics.

BRIEF CONTENTS

Preface xv

Introduction 1

Chapter 1: An Introduction to Statistical Significance 7

Chapter 2: Statistical Power and Underpowered Statistics 15

Chapter 3: Pseudoreplication: Choose Your Data Wisely 31

Chapter 4: The p Value and the Base Rate Fallacy 39

Chapter 5: Bad Judges of Significance 55

Chapter 6: Double-Dipping in the Data 63

Chapter 7: Continuity Errors 73

Chapter 8: Model Abuse 79

Chapter 9: Researcher Freedom: Good Vibrations? 89

Chapter 10: Everybody Makes Mistakes 97

Chapter 11: Hiding the Data 105

Chapter 12: What Can Be Done? 119

Notes 131

Index 147

CONTENTS IN DETAIL

PREFACE   xv
Acknowledgments xvii

INTRODUCTION   1

1
AN INTRODUCTION TO
STATISTICAL SIGNIFICANCE   7
The Power of p Values 8
       Psychic Statistics 10
       Neyman-Pearson Testing 11
Have Confidence in Intervals 12

2
STATISTICAL POWER AND
UNDERPOWERED STATISTICS   15
The Power Curve 15
The Perils of Being Underpowered 18
       Wherefore Poor Power? 20
       Wrong Turns on Red 21
Confidence Intervals and Empowerment 22
Truth Inflation 23
       Little Extremes 26

3
PSEUDOREPLICATION:
CHOOSE YOUR DATA WISELY   31
Pseudoreplication in Action 32
Accounting for Pseudoreplication 33
Batch Biology 34
Synchronized Pseudoreplication 35

4
THE P VALUE AND
THE BASE RATE FALLACY   **39**

The Base Rate Fallacy 40
    A Quick Quiz 41
    The Base Rate Fallacy in Medical Testing 42
    How to Lie with Smoking Statistics 43
    Taking Up Arms Against the Base Rate Fallacy 45
If At First You Don't Succeed, Try, Try Again 47
Red Herrings in Brain Imaging 51
Controlling the False Discovery Rate 52

5
BAD JUDGES OF SIGNIFICANCE   **55**

Insignificant Differences in Significance 55
Ogling for Significance 59

6
DOUBLE-DIPPING IN THE DATA   **63**

Circular Analysis 64
Regression to the Mean 67
Stopping Rules 68

7
CONTINUITY ERRORS   **73**

Needless Dichotomization 74
Statistical Brownout 75
Confounded Confounding 76

8
MODEL ABUSE   **79**

Fitting Data to Watermelons 80
Correlation and Causation 84
Simpson's Paradox 85

9
RESEARCHER FREEDOM:
GOOD VIBRATIONS?   **89**

A Little Freedom Is a Dangerous Thing 91
Avoiding Bias 93

10
EVERYBODY MAKES MISTAKES   97

Irreproducible Genetics 98
Making Reproducibility Easy 100
Experiment, Rinse, Repeat 102

11
HIDING THE DATA   105

Captive Data 106
    Obstacles to Sharing 107
    Data Decay 108
Just Leave Out the Details 110
    Known Unknowns............................. 110
    Outcome Reporting Bias 111
Science in a Filing Cabinet 113
    Unpublished Clinical Trials..................... 114
    Spotting Reporting Bias 115
    Forced Disclosure 116

12
WHAT CAN BE DONE?   119

Statistical Education 121
Scientific Publishing..................................... 124
Your Job.. 126

NOTES   131

INDEX   147

PREFACE

A few years ago I was an undergraduate physics major at the University of Texas at Austin. I was in a seminar course, trying to choose a topic for the 25-minute presentation all students were required to give.

"Something about conspiracy theories," I told Dr. Brent Iverson, but he wasn't satisfied with that answer. It was too broad, he said, and an engaging presentation needs to be focused and detailed. I studied the sheet of suggested topics in front of me. "How about scientific fraud and abuse?" he asked, and I agreed.

In retrospect, I'm not sure how scientific fraud and abuse is a narrower subject than conspiracy theories, but it didn't matter. After several slightly obsessive hours of research, I realized that scientific fraud isn't terribly interesting—at least, not compared to all the errors scientists commit *unintentionally*.

Woefully underqualified to discuss statistics, I nonetheless dug up several dozen research papers reporting on the numerous statistical errors routinely committed by scientists, read and outlined them, and devised a presentation that satisfied Dr. Iverson. I decided that as a future scientist (and now a self-designated statistical pundit), I should take a course in statistics.

Two years and two statistics courses later, I enrolled as a graduate student in statistics at Carnegie Mellon University. I still take obsessive pleasure in finding ways to do statistics wrong.

Statistics Done Wrong is a guide to the more egregious statistical fallacies regularly committed in the name of science. Because many scientists receive no formal statistical training— and because I do not want to limit my audience to the statistically initiated—this book assumes no formal statistical training. Some readers may easily skip through the first chapter, but I suggest at least skimming it to become familiar with my explanatory style.

My goal is not just to teach you the names of common errors and provide examples to laugh at. As much as is possible without detailed mathematics, I've explained *why* the statistical errors are errors, and I've included surveys showing how common most of these errors are. This makes for harder reading, but I think the depth is worth it. A firm understanding of basic statistics is essential for everyone in science.

For those who perform statistical analyses for their day jobs, there are "Tips" at the end of most chapters to explain what statistical techniques you might use to avoid common pitfalls. But this is not a textbook, so I will not teach you how to use these techniques in any technical detail. I hope only to make you aware of the most common problems so you are able to pick the statistical technique best suited to your question.

In case I pique your curiosity about a topic, a comprehensive bibliography is included, and every statistical misconception is accompanied by references. I omitted a great deal of mathematics in this guide in favor of conceptual understanding, but if you prefer a more rigorous treatment, I encourage you to read the original papers.

I must caution you before you read this book. Whenever we understand something that few others do, it is tempting to find every opportunity to prove it. Should *Statistics Done Wrong* miraculously become a *New York Times* best seller, I expect to see what Paul Graham calls "middlebrow dismissals" in response to any science news in the popular press. Rather than taking the time to understand the interesting parts of scientific research, armchair statisticians snipe at news articles, using the vague description of the study regurgitated from some overenthusiastic university press release to criticize the statistical design of the research.*

This already happens on most websites that discuss science news, and it would annoy me endlessly to see this book used to justify it. The first comments on a news article are always complaints about how "they didn't control for this variable" and "the sample size is too small," and 9 times out of 10, the commenter never read the scientific paper to notice that their complaint was addressed in the third paragraph.

This is stupid. A little knowledge of statistics is not an excuse to reject all of modern science. A research paper's statistical methods can be judged only in detail and in context with the rest of its methods: study design, measurement techniques, cost constraints, and goals. Use your statistical knowledge to better understand the strengths, limitations, and potential biases of research, not to shoot down any paper that seems to misuse a p value or contradict your personal beliefs. Also, remember that a conclusion supported by poor statistics can still be correct—statistical and logical errors do not make a conclusion wrong, but merely unsupported.

In short, please practice statistics responsibly. I hope you'll join me in a quest to improve the science we all rely on.

Acknowledgments

Thanks to James Scott, whose statistics courses started my statistical career and gave me the background necessary to write this book; to Raye Allen, who made James's homework assignments much more fun; to Matthew Watson and Moriel Schottlender, who gave invaluable feedback and suggestions on my drafts; to my parents, who gave suggestions and feedback; to Dr. Brent Iverson, whose seminar first motivated me to learn about statistical abuse; and to all the scientists and statisticians who have broken the rules and given me a reason to write.

My friends at Carnegie Mellon contributed many ideas and answered many questions, always patiently listening as I tried to explain some new statistical error. My professors, particularly Jing Lei, Valérie Ventura, and Howard Seltman, prepared me with the necessary knowledge. As technical reviewer, Howard

---

*Incidentally, I think this is why conspiracy theories are so popular. Once you believe you know something nobody else does (the government is out to get us!), you take every opportunity to show off that knowledge, and you end up reacting to all news with reasons why it was falsified by the government. Please don't do the same with statistical errors.

caught several embarrassing errors; if any remain, they're my responsibility, though I will claim they're merely in keeping with the title of the book.

My editors at No Starch dramatically improved the manuscript. Greg Poulos carefully read the early chapters and wasn't satisfied until he understood every concept. Leslie Shen polished my polemic in the final chapters, and the entire team made the process surprisingly easy.

I also owe thanks to the many people who emailed me suggestions and comments when the guide became available online. In no particular order, I thank Axel Boldt, Eric Franzosa, Robert O'Shea, Uri Bram, Dean Rowan, Jesse Weinstein, Peter Hozák, Chris Thorp, David Lovell, Harvey Chapman, Nathaniel Graham, Shaun Gallagher, Sara Alspaugh, Jordan Marsh, Nathan Gouwens, Arjen Noordzij, Kevin Pinto, Elizabeth Page-Gould, and David Merfield. Without their comments, my explanations would no doubt be less complete.

Perhaps you can join this list. I've tried my best, but this guide will inevitably contain errors and omissions. If you spot an error, have a question, or know a common fallacy I've missed, email me at *alex@refsmmat.com*. Any errata or updates will be published at *http://www.statisticsdonewrong.com/*.

INTRODUCTION

In the final chapter of his famous book *How to Lie with Statistics,* Darrell Huff tells us that "anything smacking of the medical profession" or backed by scientific laboratories and universities is worthy of our trust—not unconditional trust but certainly more trust than we'd afford the media or politicians. (After all, Huff's book is filled with the misleading statistical trickery used in politics and the media.) But few people complain about statistics done by trained scientists. Scientists seek understanding, not ammunition to use against political opponents.

Statistical data analysis is fundamental to science. Open a random page in your favorite medical journal and you'll be deluged with statistics: t tests, p values, proportional hazards models, propensity scores, logistic regressions, least-squares fits, and confidence intervals. Statisticians have provided scientists with tools of enormous power to find order and meaning in the most complex of datasets, and scientists have embraced them with glee.

They have not, however, embraced statistics *education*, and many undergraduate programs in the sciences require no statistical training whatsoever.

Since the 1980s, researchers have described numerous statistical fallacies and misconceptions in the popular peer-reviewed scientific literature and have found that many scientific papers—perhaps more than half—fall prey to these errors. Inadequate statistical power renders many studies incapable of finding what they're looking for, multiple comparisons and misinterpreted *p* values cause numerous false positives, flexible data analysis makes it easy to find a correlation where none exists, and inappropriate model choices bias important results. Most errors go undetected by peer reviewers and editors, who often have no specific statistical training, because few journals employ statisticians to review submissions and few papers give sufficient statistical detail to be accurately evaluated.

The problem isn't fraud but poor statistical education— poor enough that some scientists conclude that most published research findings are probably false.[1] Review articles and editorials appear regularly in leading journals, demanding higher statistical standards and tougher review, but few scientists hear their pleas, and journal-mandated standards are often ignored. Because statistical advice is scattered between frequently misleading textbooks, review articles in assorted journals, and statistical research papers difficult for scientists to understand, most scientists have no easy way to improve their statistical practice.

The methodological complexity of modern research means that scientists without extensive statistical training may not be able to understand most published research in their fields. In medicine, for example, a doctor who took one standard introductory statistics course would have sufficient knowledge to fully understand only about a fifth of research articles published in the *New England Journal of Medicine*.[2] Most doctors have even less training—many medical residents learn statistics informally through journal clubs or short courses, rather than through required courses.[3] The content that *is* taught to medical students is often poorly understood, with residents averaging less than 50% correct on tests of statistical methods commonly used in medicine.[4] Even medical school faculty with research training score less than 75% correct.

The situation is so bad that even the authors of surveys of statistical knowledge lack the necessary statistical knowledge to formulate survey questions—the numbers I just quoted are misleading because the survey of medical residents included a multiple-choice question asking residents to define a *p* value and gave four incorrect definitions as the only options.[5] We can give the authors some leeway because many introductory statistics textbooks also poorly or incorrectly define this basic concept.

When the designers of scientific studies don't employ statistics with sufficient care, they can sink years of work and thousands of dollars into research that cannot possibly answer the questions it is meant to answer. As psychologist Paul Meehl complained,

> Meanwhile our eager-beaver researcher, undismayed by logic-of-science considerations and relying blissfully on the "exactitude" of modern statistical hypothesis-testing, has produced a long publication list and been promoted to a full professorship. In terms of his contribution to the enduring body of psychological knowledge, he has done hardly anything. His true position is that of a potent-but-sterile intellectual rake, who leaves in his merry path a long train of ravished maidens but no viable scientific offspring.[6]

Perhaps it is unfair to accuse most scientists of intellectual infertility, since most scientific fields rest on more than a few misinterpreted *p* values. But these errors have massive impacts on the real world. Medical clinical trials direct our health care and determine the safety of powerful new prescription drugs, criminologists evaluate different strategies to mitigate crime, epidemiologists try to slow down new diseases, and marketers and business managers try to find the best way to sell their products—it all comes down to statistics. Statistics done wrong.

Anyone who's ever complained about doctors not making up their minds about what is good or bad for you understands the scope of the problem. We now have a dismissive attitude toward news articles claiming some food or diet or exercise might harm us—we just wait for the inevitable second study some months later, giving exactly the opposite result. As one prominent epidemiologist noted, "We are fast becoming a nuisance to society. People don't take us seriously anymore, and when they do take us seriously, we may unintentionally do more harm than good."[7] Our instincts are right. In many fields, initial results tend to be contradicted by later results. It seems the pressure to publish exciting results early and often has surpassed the responsibility to publish carefully checked results supported by a surplus of evidence.

Let's not judge so quickly, though. Some statistical errors result from a simple lack of funding or resources. Consider the mid-1970s movement to allow American drivers to turn right at red lights, saving gas and time; the evidence suggesting this would cause no more crashes than before was statistically flawed, as you will soon see, and the change cost many lives. The only factor holding back traffic safety researchers was a lack of data. Had they the money to collect more data and perform more studies—and the time to collate results from independent researchers in many different states—the truth would have been obvious.

While Hanlon's razor directs us to "never attribute to malice that which is adequately explained by incompetence," there are some published results of the "lies, damned lies, and statistics" sort. The pharmaceutical industry seems particularly tempted to bias evidence by neglecting to publish studies that show their drugs do not work;* subsequent reviewers of the literature may be pleased to find that 12 studies indicate a drug works, without knowing that 8 other unpublished studies suggest it does not. Of course, it's likely that such results would not be published by peer-reviewed journals even if they were submitted—a strong bias against unexciting results means that studies saying "it didn't work" never appear and other researchers never see them. Missing data and publication bias plague science, skewing our perceptions of important issues.

Even properly done statistics can't be trusted. The plethora of available statistical techniques and analyses grants researchers an enormous amount of freedom when analyzing their data, and it is trivially easy to "torture the data until it confesses." Just try several different analyses offered by your statistical software until one of them turns up an interesting result, and then pretend this is the analysis you intended to do all along. Without psychic powers, it's almost impossible to tell when a published result was obtained through data torture.

In "softer" fields, where theories are less quantitative, experiments are difficult to design, and methods are less standardized, this additional freedom causes noticeable biases.[8]

---

*Readers interested in the pharmaceutical industry's statistical misadventures may enjoy Ben Goldacre's *Bad Pharma* (Faber & Faber, 2012), which caused a statistically significant increase in my blood pressure while I read it.

Researchers in the United States must produce and publish interesting results to advance their careers; with intense competition for a small number of available academic jobs, scientists cannot afford to spend months or years collecting and analyzing data only to produce a statistically insignificant result. Even without malicious intent, these scientists tend to produce exaggerated results that more strongly favor their hypotheses than the data should permit.

In the coming pages, I hope to introduce you to these common errors and many others. Many of the errors are prevalent in vast swaths of the published literature, casting doubt on the findings of thousands of papers.

In recent years there have been many advocates for statistical reform, and naturally there is disagreement among them on the best method to address these problems. Some insist that p values, which I will show are frequently misleading and confusing, should be abandoned altogether; others advocate a "new statistics" based on confidence intervals. Still others suggest a switch to new Bayesian methods that give more-interpretable results, while others believe statistics as it's currently taught is just fine but used poorly. All of these positions have merits, and I am not going to pick one to advocate in this book. My focus is on statistics as it is currently used by practicing scientists.

1

AN INTRODUCTION TO STATISTICAL SIGNIFICANCE

Much of experimental science comes down to measuring differences. Does one medicine work better than another? Do cells with one version of a gene synthesize more of an enzyme than cells with another version? Does one kind of signal processing algorithm detect pulsars better than another? Is one catalyst more effective at speeding a chemical reaction than another?

We use statistics to make judgments about these kinds of differences. We will always observe *some* difference due to luck and random variation, so statisticians talk about *statistically significant* differences when the difference is larger than could easily be produced by luck. So first we must learn how to make that decision.

The Power of p Values

Suppose you're testing cold medicines. Your new medicine promises to cut the duration of cold symptoms by a day. To prove this, you find 20 patients with colds, give half of them your new medicine, and give the other half a placebo. Then you track the length of their colds and find out what the average cold length was with and without the medicine.

But not all colds are identical. Maybe the average cold lasts a week, but some last only a few days. Others might drag on for two weeks or more. It's possible that the group of 10 patients who got the genuine medicine in your study all came down with really short colds. How can you prove that your medicine works, rather than just proving that some patients got lucky?

Statistical hypothesis testing provides the answer. If you know the distribution of typical cold cases—roughly how many patients get short colds, long colds, and average-length colds—you can tell how likely it is that a random sample of patients will all have longer or shorter colds than average. By performing a *hypothesis test* (also known as a *significance test*), you can answer this question: "Even if my medication were completely ineffective, what are the chances my experiment would have produced the observed outcome?"

If you test your medication on only one person, it's not too surprising if her cold ends up being a little shorter than usual. Most colds aren't perfectly average. But if you test the medication on 10 million patients, it's pretty unlikely that all those patients will just happen to get shorter colds. More likely, your medication actually works.

Scientists quantify this intuition with a concept called the *p value*. The *p* value is the probability, under the assumption that there is no true effect or no true difference, of collecting data that shows a difference equal to or more extreme than what you actually observed.

So if you give your medication to 100 patients and find that their colds were a day shorter on average, then the *p* value of this result is the chance that if your medication didn't actually do anything, their average cold would be a day shorter than the control group's by luck alone. As you might guess, the *p* value depends on the size of the effect—colds that are shorter by four days are less common than colds that are shorter by just one day—as well as on the number of patients you test the medication on.

Remember, a *p* value is not a measure of how right you are or how important a difference is. Instead, think of it as a measure of surprise. If you assume your medication is ineffective and there is no reason other than luck for the two groups to differ, then the smaller the *p* value, the more surprising and lucky your results are—or your assumption is wrong, and the medication truly works.

How do you translate a *p* value into an answer to this question: "Is there really a difference between these groups?" A common rule of thumb is to say that any difference where $p < 0.05$ is statistically significant. The choice of 0.05 isn't because of any special logical or statistical reasons, but it has become scientific convention through decades of common use.

Notice that the *p* value works by assuming there is no difference between your experimental groups. This is a counter-intuitive feature of significance testing: if you want to prove that your drug works, you do so by showing the data is *in*consistent with the drug *not* working. Because of this, *p* values can be extended to any situation where you can mathematically express a hypothesis you want to knock down.

But *p* values have their limitations. Remember, *p* is a measure of surprise, with a smaller value suggesting that you should be more surprised. It's not a measure of the size of the effect. You can get a tiny *p* value by measuring a huge effect—"This medicine makes people live four times longer"—or by measuring a tiny effect with great certainty. And because any medication or intervention usually has *some* real effect, you can always get a statistically significant result by collecting so much data that you detect extremely tiny but relatively unimportant differences. As Bruce Thompson wrote,

> Statistical significance testing can involve a tautological logic in which tired researchers, having collected data on hundreds of subjects, then conduct a statistical test to evaluate whether there were a lot of subjects, which the researchers already know, because they collected the data and know they are tired. This tautology has created considerable damage as regards the cumulation of knowledge.[1]

In short, statistical significance does not mean your result has any *practical* significance. As for statistical *in*significance, it doesn't tell you much. A statistically insignificant difference could be nothing but noise, or it could represent a real effect that can be pinned down only with more data.

There's no mathematical tool to tell you whether your hypothesis is true or false; you can see only whether it's consistent with the data. If the data is sparse or unclear, your conclusions will be uncertain.

Psychic Statistics

Hidden beneath their limitations are some subtler issues with p values. Recall that a p value is calculated under the assumption that luck (not your medication or intervention) is the only factor in your experiment, and that p is defined as the probability of obtaining a result equal to *or more extreme* than the one observed. This means p values force you to reason about results that never actually occurred—that is, results more extreme than yours. The probability of obtaining such results depends on your experimental design, which makes p values "psychic": two experiments with different designs can produce identical data but different p values because the *unobserved* data is different.

Suppose I ask you a series of 12 true-or-false questions about statistical inference, and you correctly answer 9 of them. I want to test the hypothesis that you answered the questions by guessing randomly. To do this, I need to compute the chances of you getting *at least* 9 answers right by simply picking true or false randomly for each question. Assuming you pick true and false with equal probability, I compute $p = 0.073$.* And since $p > 0.05$, it's plausible that you guessed randomly. If you did, you'd get 9 or more questions correct 7.3% of the time.[2]

But perhaps it was not my original plan to ask you only 12 questions. Maybe I had a computer that generated a limitless supply of questions and simply asked questions until you got 3 wrong. Now I have to compute the probability of you getting 3 questions wrong after being asked 15 or 20 or 47 of them. I even have to include the remote possibility that you made it to 175,231 questions before getting 3 questions wrong. Doing the math, I find that $p = 0.033$. Since $p < 0.05$, I conclude that random guessing would be unlikely to produce this result.

This is troubling: two experiments can collect identical data but result in different conclusions. Somehow, the p value can read your intentions.

---

*I used a probability distribution known as the *binomial distribution* to calculate this result. In the next paragraph, I'll calculate p using a different distribution, called the *negative binomial distribution*. A detailed explanation of probability distributions is beyond the scope of this book; we're more interested in how to interpret p values rather than how to calculate them.

Neyman-Pearson Testing

To better understand the problems of the p value, you need to learn a bit about the history of statistics. There are two major schools of thought in statistical significance testing. The first was popularized by R.A. Fisher in the 1920s. Fisher viewed p as a handy, informal method to see how surprising a set of data might be, rather than part of some strict formal procedure for testing hypotheses. The p value, when combined with an experimenter's prior experience and domain knowledge, could be useful in deciding how to interpret new data.

After Fisher's work was introduced, Jerzy Neyman and Egon Pearson tackled some unanswered questions. For example, in the cold medicine test, you can choose to compare the two groups by their means, medians, or whatever other formula you might concoct, so long as you can derive a p value for the comparison. But how do you know which is best? What does "best" even mean for hypothesis testing?

In science, it is important to limit two kinds of errors: *false positives*, where you conclude there is an effect when there isn't, and *false negatives*, where you fail to notice a real effect. In some sense, false positives and false negatives are flip sides of the same coin. If we're too ready to jump to conclusions about effects, we're prone to get false positives; if we're too conservative, we'll err on the side of false negatives.

Neyman and Pearson reasoned that although it's impossible to eliminate false positives and negatives entirely, it *is* possible to develop a formal decision-making process that will ensure false positives occur only at some predefined rate. They called this rate α, and their idea was for experimenters to set an α based upon their experience and expectations. So, for instance, if we're willing to put up with a 10% rate of false positives, we'll set $\alpha = 0.1$. But if we need to be more conservative in our judgments, we might set α at 0.01 or lower. To determine which testing procedure is best, we see which has the lowest false negative rate for a given choice of α.

How does this work in practice? Under the Neyman–Pearson system, we define a *null hypothesis*—a hypothesis that there is no effect—as well as an *alternative hypothesis*, such as "The effect is greater than zero." Then we construct a test that compares the two hypotheses, and determine what results we'd expect to see were the null hypothesis true. We use the p value to implement the Neyman-Pearson testing procedure by rejecting the null hypothesis whenever $p < \alpha$. Unlike Fisher's procedure, this method deliberately does not address the strength of evidence in any one particular experiment; now we are interested in only the decision to reject or not. The size of the p value isn't used to compare experiments or draw any conclusions besides "The null hypothesis can be rejected." As Neyman and Pearson wrote,

> We are inclined to think that as far as a particular hypothesis is concerned, no test based upon the theory of probability can by itself provide any valuable evidence of the truth or falsehood of that hypothesis.
>
> But we may look at the purpose of tests from another view-point. Without hoping to know whether each separate hypothesis is true or false, we may search for rules to govern our behaviour with regard to them, in following which we insure that, in the long run of experience, we shall not be too often wrong.[3]

Although Neyman and Pearson's approach is conceptually distinct from Fisher's, practicing scientists often conflate the two.[4,5,6] The Neyman-Pearson approach is where we get "statistical significance," with a prechosen p value threshold that guarantees the long-run false positive rate. But suppose you run an experiment and obtain $p = 0.032$. If your threshold was the conventional $p < 0.05$, this is statistically significant. But it'd also have been statistically significant if your threshold was $p < 0.033$. So it's tempting—and a common misinterpretation—to say "My false positive rate is 3.2%."

But that doesn't make sense. A single experiment does not have a false positive rate. The false positive rate is determined by your *procedure*, not the result of any single experiment. You can't claim each experiment had a false positive rate of exactly p, whatever that turned out to be, when you were using a procedure to get a long-run false positive rate of α.

Have Confidence in Intervals

Significance tests tend to receive lots of attention, with the phrase "statistically significant" now part of the popular lexicon. Research results, especially in the biological and social sciences, are commonly presented with p values. But p isn't the only way to evaluate the weight of evidence. *Confidence intervals* can answer the same questions as p values, with the advantage that they provide more information and are more straightforward to interpret.

A confidence interval combines a point estimate with the uncertainty in that estimate. For instance, you might say your new experimental drug reduces the average length of a cold by 36 hours and give a 95% confidence interval between 24 and 48 hours. (The confidence interval is for the *average* length; individual patients may have wildly varying cold lengths.) If you run 100 identical experiments, about 95 of the confidence intervals will include the true value you're trying to measure.

A confidence interval quantifies the uncertainty in your conclusions, providing vastly more information than a *p* value, which says nothing about effect sizes. If you want to test whether an effect is significantly different from zero, you can construct a 95% confidence interval and check whether the interval includes zero. In the process, you get the added bonus of learning how precise your estimate is. If the confidence interval is too wide, you may need to collect more data.

For example, if you run a clinical trial, you might produce a confidence interval indicating that your drug reduces symptoms by somewhere between 15 and 25 percent. This effect is statistically significant because the interval doesn't include zero, and now you can assess the importance of this difference using your clinical knowledge of the disease in question. As when you were using *p* values, this step is important—you shouldn't trumpet this result as a major discovery without evaluating it in context. If the symptom is already pretty innocuous, maybe a 15–25% improvement isn't too important. Then again, for a symptom like spontaneous human combustion, you might get excited about *any* improvement.

If you can write a result as a confidence interval instead of as a *p* value, you should.[7] Confidence intervals sidestep most of the interpretational subtleties associated with *p* values, making the resulting research that much clearer. So why are confidence intervals so unpopular? In experimental psychology research journals, 97% of research papers involve significance testing, but only about 10% ever report confidence intervals—and most of those don't use the intervals as supporting evidence for their conclusions, relying instead on significance tests.[8] Even the prestigious journal *Nature* falls short: 89% of its articles report *p* values without any confidence intervals or effect sizes, making their results impossible to interpret in context.[9] One journal editor noted that "*p* values are like mosquitoes" in that they "have an evolutionary niche somewhere and [unfortunately] no amount of scratching, swatting or spraying will dislodge them."[10]

One possible explanation is that confidence intervals go unreported because they are often embarrassingly wide.[11] Another is that the peer pressure of peer-reviewed science is too strong—it's best to do statistics the same way everyone else does, or else the reviewers might reject your paper. Or maybe the widespread confusion about *p* values obscures the benefits of confidence intervals. Or the overemphasis on hypothesis testing in statistics courses means most scientists don't know how to calculate and use confidence intervals.

Journal editors have sometimes attempted to enforce the reporting of confidence intervals. Kenneth Rothman, an associate editor at the *American Journal of Public Health* in the mid-1980s, began returning submissions with strongly worded letters:

> All references to statistical hypothesis testing and statistical significance should be removed from the paper. I ask that you delete *p* values as well as comments about statistical significance. If you do not agree with my standards (concerning the inappropriateness of significance tests), you should feel free to argue the point, or simply ignore what you may consider to be my misguided view, by publishing elsewhere.[12]

During Rothman's three-year tenure as associate editor, the fraction of papers reporting solely *p* values dropped precipitously. Significance tests returned after his departure, although subsequent editors successfully encouraged researchers to report confidence intervals as well. But despite reporting confidence intervals, few researchers discussed them in their articles or used them to draw conclusions, preferring instead to treat them merely as significance tests.[12]

Rothman went on to found the journal *Epidemiology*, which had a strong statistical reporting policy. Early on, authors familiar with significance testing preferred to report *p* values alongside confidence intervals, but after 10 years, attitudes had changed, and reporting only confidence intervals became common practice.[12]

Perhaps brave (and patient) journal editors can follow Rothman's example and change statistical practices in their fields.

2

STATISTICAL POWER AND UNDERPOWERED STATISTICS

You've seen how it's possible to miss real effects by not collecting enough data. You might miss a viable medicine or fail to notice an important side effect. So how do you know how much data to collect?

The concept of *statistical power* provides the answer. The power of a study is the probability that it will distinguish an effect of a certain size from pure luck. A study might easily detect a huge benefit from a medication, but detecting a subtle difference is much less likely.

The Power Curve

Suppose I'm convinced that my archnemesis has an unfair coin. Rather than getting heads half the time and tails half the time, it's biased to give one outcome 60% of the time, allowing him to cheat at incredibly boring coin-flipping betting games. I suspect he's cheating—but how to *prove* it?

I can't just take the coin, flip it 100 times, and count the heads. Even a perfectly fair coin won't always get 50 heads, as the solid line in Figure 2-1 shows.

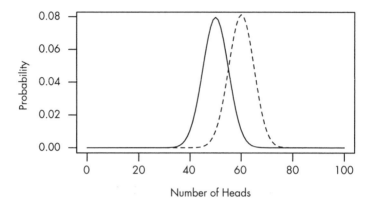

Figure 2-1: The probability of getting different numbers of heads if you flip a fair coin (solid line) or biased coin (dashed line) 100 times. The biased coin gives heads 60% of the time.

Even though 50 heads is the most likely outcome, it still happens less than 10% of the time. I'm also reasonably likely to get 51 or 52 heads. In fact, when flipping a fair coin 100 times, I'll get between 40 and 60 heads 95% of the time. On the other hand, results far outside this range are unlikely: with a fair coin, there's only a 1% chance of obtaining more than 63 or fewer than 37 heads. Getting 90 or 100 heads is almost impossible.

Compare this to the dashed line in Figure 2-1, showing the probability of outcomes for a coin biased to give heads 60% of the time. The curves do overlap, but you can see that an unfair coin is much more likely to produce 70 heads than a fair coin is.

Let's work out the math. Say I run 100 trials and count the number of heads. If the result isn't exactly 50 heads, I'll calculate the probability that a *fair* coin would have turned up a deviation of that size or larger. That probability is my p value. I'll consider a p value of 0.05 or less to be statistically significant and hence call the coin unfair if p is smaller than 0.05.

How likely am I to find out a coin is biased using this procedure? A *power curve*, as shown in Figure 2-2, can tell me. Along the horizontal axis is the coin's true probability of getting heads—that is, how biased it is. On the vertical axis is the probability that I will conclude the coin is rigged.

The *power* for any hypothesis test is the probability that it will yield a statistically significant outcome (defined in this example as $p < 0.05$). A fair coin will show between 40 and 60 heads in 95% of trials, so for an *unfair* coin, the power is the probability of a result *outside* this range of 40–60 heads. The power is affected by three factors:

- **The size of the bias you're looking for.** A huge bias is much easier to detect than a tiny one.

- **The sample size.** By collecting more data (more coin flips), you can more easily detect small biases.

- **Measurement error.** It's easy to count coin flips, but many experiments deal with values that are harder to measure, such as medical studies investigating symptoms of fatigue or depression.

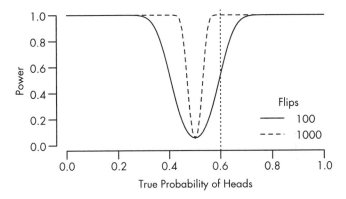

Figure 2-2: The power curves for 100 and 1,000 coin flips, showing the probability of detecting biases of different magnitudes. The vertical line indicates a 60% probability of heads.

Let's start with the size of the bias. The solid line in Figure 2-2 shows that if the coin is rigged to give heads 60% of the time, I have a 50% chance of concluding that it's rigged after 100 flips. (That is, when the true probability of heads is 0.6, the power is 0.5.) The other half of the time, I'll get fewer than 60 heads and fail to detect the bias. With only 100 flips, there's just too little data to *always* separate bias from random variation. The coin would have to be incredibly biased—yielding heads more than 80% of the time, for example—for me to notice nearly 100% of the time.

Another problem is that even if the coin is perfectly fair, I will falsely accuse it of bias 5% of the time. I've designed my test to interpret outcomes with $p < 0.05$ as a sign of bias, but those outcomes *do* happen even with a fair coin.

Fortunately, an increased sample size improves the sensitivity. The dashed line shows that with 1,000 flips, I can easily tell whether the coin is rigged. This makes sense: it's overwhelmingly unlikely that I could flip a fair coin 1,000 times and get more than 600 heads. I'll get between 469 and 531 95% of the time. Unfortunately, I don't really have the time to flip my nemesis's coin 1,000 times to test its fairness. Often, performing a sufficiently powerful test is out of the question for purely practical reasons.

Now counting heads and tails is easy, but what if I were instead administering IQ tests? An IQ score does not measure an underlying "truth" but instead can vary from day to day depending on the questions on the test and the mood of the subject, introducing random noise to the measurements. If you were to compare the IQs of two groups of people, you'd see not only the normal variation in intelligence from one person to the next but also the random variation in *individual* scores. A test with high variability, such as an IQ test requiring subjective grading, will have relatively less statistical power.

More data helps distinguish the signal from the noise. But this is easier said than done: many scientists don't have the resources to conduct studies with adequate statistical power to detect what they're looking for. They are doomed to fail before they even start.

The Perils of Being Underpowered

Consider a trial testing two different medicines, Fixitol and Solvix, for the same condition. You want to know which is safer, but side effects are rare, so even if you test both medicines on 100 patients, only a few in each group will suffer serious side effects. Just as it is difficult to tell the difference between two coins that turn up 50% heads and 51% heads, the difference between a 3% and 4% side effect rate is difficult to discern. If four people taking Fixitol have serious side effects and only three people taking Solvix have them, you can't say for sure whether the difference is due to Fixitol.

If a trial isn't powerful enough to detect the effect it's looking for, we say it is *underpowered.*

You might think calculations of statistical power are essential for medical trials; a scientist might want to know how many patients are needed to test a new medication, and a quick calculation of statistical power would provide the answer. Scientists are usually satisfied when the statistical power is 0.8 or higher, corresponding to an 80% chance of detecting a real effect of the expected size. (If the true effect is actually larger, the study will have greater power.)

However, few scientists ever perform this calculation, and few journal articles even mention statistical power. In the prestigious journals *Science* and *Nature*, fewer than 3% of articles calculate statistical power before starting their study.[1] Indeed, many trials conclude that "there was no statistically significant difference in adverse effects between groups," without noting that there was insufficient data to detect any but the largest differences.[2] If one of these trials was comparing side effects in two drugs, a doctor might erroneously think the medications are equally safe, when one could very well be much more dangerous than the other.

Maybe this is a problem only for rare side effects or only when a medication has a weak effect? Nope. In one sample of studies published in prestigious medical journals between 1975 and 1990, more than four-fifths of randomized controlled trials that reported negative results didn't collect enough data to detect a *25% difference* in primary outcome between treatment groups. That is, even if one medication reduced symptoms by 25% more than another, there was insufficient data to make that conclusion. And nearly *two-thirds* of the negative trials didn't have the power to detect a 50% difference.[3]

A more recent study of trials in cancer research found similar results: only about half of published studies with negative results had enough statistical power to detect even a large difference in their primary outcome variable.[4] Less than 10% of these studies explained why their sample sizes were so poor. Similar problems have been consistently seen in other fields of medicine.[5,6]

In neuroscience, the problem is even worse. Each individual neuroscience study collects such little data that the median study has only a 20% chance of being able to detect the effect it's looking for. You could compensate for this by aggregating data collected across several papers all investigating the same effect. But since many neuroscience studies use animal subjects, this raises a significant ethical concern. If each study is underpowered, the true effect will likely be discovered only after many studies using many animals have been completed and analyzed—using far more animal subjects than if the study had been done properly in the first place.[7] An ethical review board should not approve a trial if it knows the trial is unable to detect the effect it is looking for.

Wherefore Poor Power?

Curiously, the problem of underpowered studies has been known for decades, yet it is as prevalent now as it was when first pointed out. In 1960 Jacob Cohen investigated the statistical power of studies published in the *Journal of Abnormal and Social Psychology*[8] and discovered that the average study had only a power of 0.48 for detecting medium-sized effects.* His research was cited hundreds of times, and many similar reviews followed, all exhorting the need for power calculations and larger sample sizes. Then, in 1989, a review showed that in the decades since Cohen's research, the average study's power had actually *decreased*.[9] This decrease was because of researchers becoming aware of another problem, the issue of multiple comparisons, and compensating for it in a way that reduced their studies' power. (I will discuss multiple comparisons in Chapter 4, where you will see that there is an unfortunate trade-off between a study's power and multiple comparison correction.)

So why are power calculations often forgotten? One reason is the discrepancy between our intuitive feeling about sample sizes and the results of power calculations. It's easy to think, "Surely these are enough test subjects," even when the study has abysmal power. For example, suppose you're testing a new heart attack treatment protocol and hope to cut the risk of death in half, from 20% to 10%. You might be inclined to think, "If I don't see a difference when I try this procedure on 50 patients, clearly the benefit is too small to be useful." But to have 80% power to detect the effect, you'd actually need *400* patients—200 in each control and treatment group.[10] Perhaps clinicians just don't realize that their adequate-seeming sample sizes are in fact far too small.

Math is another possible explanation for why power calculations are so uncommon: analytically calculating power can be difficult or downright impossible. Techniques for calculating power are not frequently taught in intro statistics courses. And some commercially available statistical software does not come with power calculation functions. It is possible to avoid hairy mathematics by simply simulating thousands of artificial datasets with the effect size you expect and running your statistical tests on the simulated data. The power is simply the fraction of datasets for which you obtain a statistically significant result. But this approach requires programming experience, and simulating realistic data can be tricky.

---

*Cohen defined "medium-sized" as a 0.5-standard-deviation difference between groups.

Even so, you'd think scientists would notice their power problems and try to correct them; after five or six studies with insignificant results, a scientist might start wondering what she's doing wrong. But the average study performs not one hypothesis test but many and so has a good shot at finding *something* significant.[11] As long as this significant result is interesting enough to feature in a paper, the scientist will not feel that her studies are underpowered.

The perils of insufficient power do not mean that scientists are lying when they state they detected no significant difference between groups. But it's misleading to assume these results mean there is no *real* difference. There may be a difference, even an important one, but the study was so small it'd be lucky to notice it. Let's consider an example we see every day.

Wrong Turns on Red

In the 1970s, many parts of the United States began allowing drivers to turn right at a red light. For many years prior, road designers and civil engineers argued that allowing right turns on a red light would be a safety hazard, causing many additional crashes and pedestrian deaths. But the 1973 oil crisis and its fallout spurred traffic agencies to consider allowing right turns on red to save fuel wasted by commuters waiting at red lights, and eventually Congress required states to allow right turns on red, treating it as an energy conservation measure just like building insulation standards and more efficient lighting.

Several studies were conducted to consider the safety impact of the change. In one, a consultant for the Virginia Department of Highways and Transportation conducted a before-and-after study of 20 intersections that had begun to allow right turns on red. Before the change, there were 308 accidents at the intersections; after, there were 337 in a similar length of time. But this difference was not statistically significant, which the consultant indicated in his report. When the report was forwarded to the governor, the commissioner of the Department of Highways and Transportation wrote that "we can discern no significant hazard to motorists or pedestrians from implementation" of right turns on red.[12] In other words, he turned *statistical* insignificance into *practical* insignificance.

Several subsequent studies had similar findings: small increases in the number of crashes but not enough data to conclude these increases were significant. As one report concluded,

> There is no reason to suspect that pedestrian accidents involving RT operations (right turns) have increased after the adoption of [right turn on red].

Of course, these studies were underpowered. But more cities and states began to allow right turns on red, and the practice became widespread across the entire United States. Apparently, no one attempted to aggregate these many small studies to produce a more useful dataset. Meanwhile, more pedestrians were being run over, and more cars were involved in collisions. Nobody collected enough data to show this conclusively until several years later, when studies finally showed that among incidents involving right turns, collisions were occurring roughly 20% more frequently, 60% more pedestrians were being run over, and twice as many bicyclists were being struck.[13,14,*]

Alas, the world of traffic safety has learned little from this example. A 2002 study, for example, considered the impact of paved shoulders on the accident rates of traffic on rural roads. Unsurprisingly, a paved shoulder reduced the risk of accident—but there was insufficient data to declare this reduction statistically significant, so the authors stated that the cost of paved shoulders was not justified. They performed no cost-benefit analysis because they treated the insignificant difference as meaning there was no difference at all, despite the fact that they had collected data suggesting that paved shoulders improved safety! The evidence was not strong enough to meet their desired p value threshold.[12] A better analysis would have admitted that while it is plausible that shoulders have no benefit at all, the data is also consistent with them having substantial benefits. That means looking at *confidence intervals*.

Confidence Intervals and Empowerment

More useful than a statement that an experiment's results were statistically insignificant is a confidence interval giving plausible sizes for the effect. Even if the confidence interval includes zero, its width tells you a lot: a narrow interval covering zero tells you that the effect is most likely small (which may be all you need to know, if a small effect is not practically useful),

---

*It is important to note that accidents involving right turns are rare: these changes amount to fewer than 100 deaths per year in the United States.[15] A 60% increase in a small number is still small—but nonetheless, a statistical error kills dozens of people each year!

while a wide interval clearly shows that the measurement was not precise enough to draw conclusions.

Physicists commonly use confidence intervals to place bounds on quantities that are not significantly different from zero. In the search for a new fundamental particle, for example, it's not helpful to say, "The signal was not statistically significant." Instead, physicists can use a confidence interval to place an upper bound on the rate at which the particle is produced in the particle collisions under study and then compare this result to the competing theories that predict its behavior (and force future experimenters to build yet bigger instruments to find it).

Thinking about results in terms of confidence intervals provides a new way to approach experimental design. Instead of focusing on the power of significance tests, ask, "How much data must I collect to measure the effect to my desired precision?" Even a powerful experiment can nonetheless produce significant results with extremely wide confidence intervals, making its results difficult to interpret.

Of course, the sizes of our confidence intervals vary from one experiment to the next because our data varies from experiment to experiment. Instead of choosing a sample size to achieve a certain level of power, we choose a sample size so the confidence interval will be suitably narrow 99% of the time (or 95%; there's not yet a standard convention for this number, called the *assurance*, which determines how often the confidence interval must beat our target width).[16]

Sample size selection methods based on assurance have been developed for many common statistical tests, though not for all; it is a new field, and statisticians have yet to fully explore it.[17] (These methods go by the name *accuracy in parameter estimation*, or *AIPE*.) Statistical power is used far more often than assurance, which has not yet been widely adopted by scientists in any field. Nonetheless, these methods are enormously useful. Statistical significance is often a crutch, a catchier-sounding but less informative substitute for a good confidence interval.

Truth Inflation

Suppose Fixitol reduces symptoms by 20% over a placebo, but the trial you're using to test it is too small to have adequate statistical power to detect this difference reliably. We know that small trials tend to have varying results; it's easy to get 10 lucky patients who have shorter colds than usual but much harder to get 10,000 who all do.

Now imagine running many copies of this trial. Sometimes you get unlucky patients, so you don't notice any statistically significant improvement from your drug. Sometimes your patients are exactly average and the treatment group has their symptoms reduced by 20%—but you don't have enough data to call this a statistically significant increase, so you ignore it. Sometimes the patients are lucky and have their symptoms reduced by much more than 20%, so you stop the trial and say, "Look! It works!" You can plot these outcomes in Figure 2-3, which shows the probability that each trial will yield a certain effect size estimate.

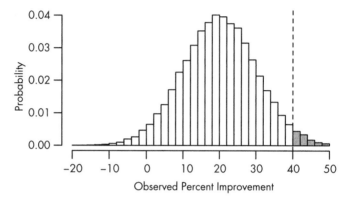

Figure 2-3: If you run your trial thousands of times, you will see a broad distribution of effect sizes in terms of percent reduction in symptoms. The vertical dotted line indicates the effect size which is large enough to be statistically significant. The true improvement is 20%, but you see effects from 10% losses to 50% gains. Only the lucky trials are statistically significant, exaggerating the effect size.

You've correctly concluded Fixitol is effective, but you've inflated the size of its effect because your study was underpowered.

This effect, known as *truth inflation*, *type M error* (*M* for magnitude), or the *winner's curse*, occurs in fields where many researchers conduct similar experiments and compete to publish the most "exciting" results: pharmacological trials, epidemiological studies, gene association studies ("gene A causes condition B"), and psychological studies often show symptoms, along with some of the most-cited papers in the medical literature.[18,19] In fast-moving fields such as genetics, the earliest published results are often the most extreme because journals are most interested in publishing new and exciting results. Follow-up studies tend to show much smaller effects.[20]

Consider also that top-ranked journals, such as *Nature* and *Science*, prefer to publish studies with groundbreaking results—meaning large effect sizes in novel fields with little prior research. This is a perfect combination for chronic truth inflation. Some evidence suggests a correlation between a journal's impact factor (a rough measure of its prominence and importance) and the factor by which its studies overestimate effect sizes. Studies that produce less "exciting" results are closer to the truth but less interesting to a major journal editor.[21,22]

When a study claims to have detected a large effect with a relatively small sample, your first reaction should not be "Wow, they've found something big!" but "Wow, this study is underpowered!"[23] Here's an example. Starting in 2005, Satoshi Kanazawa published a series of papers on the theme of gender ratios, culminating with "Beautiful Parents Have More Daughters." He followed up with a book discussing this and other "politically incorrect truths" he'd discovered. The studies were popular in the press at the time, particularly because of the large effect size they reported: Kanazawa claimed the most beautiful parents have daughters 52% of the time, but the least attractive parents have daughters only 44% of the time.

To biologists, a small effect—perhaps one or two percentage points—would be plausible. The *Trivers–Willard Hypothesis* suggests that if parents have a trait that benefits girls more than boys, then they will have more girls than boys (or vice versa). If you assume girls benefit more from beauty than boys, then the hypothesis would predict beautiful parents would have, on average, slightly more daughters.

But the effect size claimed by Kanazawa was extraordinary. And as it turned out, he committed several errors in his statistical analysis. A corrected regression analysis found that his data showed attractive parents were indeed 4.7% more likely to have girls—but the confidence interval stretched from 13.3% more likely to 3.9% *less* likely.[23] Though Kanazawa's study used data from nearly 3,000 parents, the results were not statistically significant.

Enormous amounts of data would be needed to reliably detect a small difference. Imagine a more realistic effect size—say, 0.3%. Even with 3,000 parents, an observed difference of 0.3% is far too small to distinguish from luck. You'd be lucky to obtain a statistically significant result just 5% of the time. These results will be those that exaggerate the true effect by at least a factor of 20, and 40% of them will produce a wild overestimate in favor of boys instead of girls.[23]

So even if Kanazawa had performed a perfect statistical analysis, he still would have occasionally gotten lucky with a paper like "Engineers Have More Sons, Nurses Have More Daughters"* and given a wild overestimate of a true, tiny effect. Studies of the size he conducted are simply *incapable* of detecting effects of the size you'd expect in advance. A prior power analysis would have told him this.

Little Extremes

Truth inflation arises because small, underpowered studies have widely varying results. Occasionally you're bound to get lucky and have a statistically significant but wildly overestimated result. But this wide variation can cause trouble even when you're not performing significance tests. Suppose you're in charge of public school reform. As part of your research into the best teaching methods, you look at the effect of school size on standardized test scores. Do smaller schools perform better than larger schools? Should you try to build many small schools or a few large schools?

To answer this question, you compile a list of the highest-performing schools you have. The average school has about 1,000 students, but the top-scoring 10 schools are almost all smaller than that. It seems that small schools do the best, perhaps because teachers can get to know students and help them individually.

Then you take a look at the worst-performing schools, expecting them to be large urban schools with thousands of students and overworked teachers. Surprise! They're all small schools too.

What's going on? Well, take a look at the plot of test scores versus school size in Figure 2-4. Smaller schools have wider variation in test scores because they have fewer students. With fewer students, there are fewer data points to establish the "true" performance of the teachers; a few anomalous scores can sway the school's average significantly. As schools get larger, test scores vary less and in fact *increase* on average.[24]

---

*A real paper, which he published in 2005 in the *Journal of Theoretical Biology.*

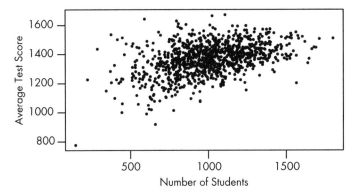

Figure 2-4: Schools with more students have less random variation in their test scores. This data is simulated but based on real observations of Pennsylvania public schools.

Another example: in the United States, counties with the lowest rates of kidney cancer tend to be Midwestern, Southern, and Western rural counties. Why might this be? Maybe rural people get more exercise or inhale less-polluted air. Or perhaps they just lead less stressful lives.

On the other hand, counties with the *highest* rates of kidney cancer tend to be Midwestern, Southern, and Western rural counties.

The problem, of course, is that rural counties have the smallest populations. A single kidney cancer patient in a county with 10 residents gives that county the highest kidney cancer rate in the nation. Small counties hence have much more variation in kidney cancer rates simply because they have so few residents.[25] The confidence intervals for their cancer rates are correspondingly larger.

A popular strategy to fight this problem is called *shrinkage*. For counties with few residents, you can "shrink" the cancer rate estimates toward the national average by taking a weighted average of the county cancer rate with the national average rate. When the county has few residents, you weight the national average strongly; when the county is large, you weight the county strongly. Shrinkage is now common practice in constructing cancer rate maps, among other applications.* Unfortunately, it biases results in the opposite direction: small

---

*However, shrinkage is usually implemented using more sophisticated methods than a simple weighted average.

counties with truly abnormal cancer rates are estimated to have rates much closer to the national average than they are.

There's no single fix to this problem. The best alternative is to sidestep it altogether: rather than estimating rates by county, you could use congressional districts, which in the United States are designed to have roughly equal populations. Congressional districts are much larger than counties, though, and frequently they have strange shapes because of gerrymandering. Maps based on districts may not be statistically misleading but are still difficult to interpret.

Of course, enforcing equal sample sizes isn't always an option. Online shopping sites, for instance, need to sort products based on customer ratings, but they can't force equal numbers of customers to rate every product. Another example is a discussion website like reddit, which can sort comments by user ratings; comments can receive vastly different numbers of votes depending on when or where or by whom they were posted. Shrinkage is helpful in dealing with these situations. An online store can use a weighted average of a product's ratings and some global average. Products with few ratings will be treated as generically average, while products with thousands of votes are sorted by their true individual ratings.

For sites like reddit that have simple up-and-down votes rather than star ratings, one alternative is to generate a confidence interval for the fraction of positive votes. The interval starts wide when a comment has only a few votes and narrows to a definite value ("70% of voters like this comment") as comments accumulate; sort the comments by the bottom bound of their confidence intervals. New comments start near the bottom, but the best among them accumulate votes and creep up the page as the confidence interval narrows. And because comments are sorted by the proportion of positive votes rather than the total number, new comments can compete with those that have already accumulated thousands of votes.[26,27]

TIPS • Calculate the statistical power when designing your study to determine the appropriate sample size. Don't skimp. Consult a book like Cohen's classic *Statistical Power Analysis for the Behavioral Sciences* or talk to a statistical consultant. If the sample size is impractical, be aware of the limitations of your study.

• When you need to measure an effect with precision, rather than simply testing for significance, use assurance instead of power: design your experiment to measure the hypothesized effect to your desired level of precision.

- Remember that "statistically insignificant" does not mean "zero." Even if your result is insignificant, it represents the best available estimate given the data you have collected. "Not significant" does not mean "nonexistent."

- Look skeptically on the results of clearly underpowered studies. They may be exaggerated due to truth inflation.

- Use confidence intervals to determine the range of answers consistent with your data, regardless of statistical significance.

- When comparing groups of different sizes, compute confidence intervals. These will reflect the additional certainty you have in larger groups.

3

PSEUDOREPLICATION: CHOOSE YOUR DATA WISELY

In a randomized controlled trial, test subjects are assigned to either experimental or control groups randomly, rather than for any systematic reason. Though the word *random* makes such studies sound slightly unscientific, a medical trial is not usually considered definitive unless it is a randomized controlled trial. Why? What's so important about randomization?

Randomization prevents researchers from introducing systematic biases between test groups. Otherwise, they might assign frail patients to a less risky or less demanding treatment or assign wealthier patients to the new treatment because their insurance companies will pay for it. But randomization has no hidden biases, and it guarantees that each group has roughly the same demographics; any confounding factors—even ones you don't know about—can't affect your results. When you obtain a statistically significant result, you know that the only possible cause is your medication or intervention.

Pseudoreplication in Action

Let me return to a medical example. I want to compare two blood pressure medications, so I recruit 2,000 patients and randomly split them into two groups. Then I administer the medications. After waiting a month for the medication to take effect, I measure each patient's blood pressure and compare the groups to find which has the lower average blood pressure. I can do an ordinary hypothesis test and get an ordinary *p* value; with my sample size of 1,000 patients per group, I will have good statistical power to detect differences between the medications.

Now imagine an alternative experimental design. Instead of 1,000 patients per group, I recruit only 10, but I measure each patient's blood pressure 100 times over the course of a few months. This way I can get a more accurate fix on their individual blood pressures, which may vary from day to day. Or perhaps I'm worried that my sphygmomanometers are not perfectly calibrated, so I measure with a different one each day.* I still have 1,000 data points per group but only 10 unique patients. I can perform the same hypothesis tests with the same statistical power since I seem to have the same sample size.

But do I really? A large sample size is *supposed* to ensure that any differences between groups are a result of my treatment, not genetics or preexisting conditions. But in this new design, I'm not recruiting new patients. I'm just counting the genetics of each existing patient 100 times.

This problem is known as *pseudoreplication*, and it is quite common.[1] For instance, after testing cells from a culture, a biologist might "replicate" his results by testing more cells from the same culture. Or a neuroscientist might test multiple neurons from the same animal, claiming to have a large sample size of hundreds of neurons from just two rats. A marine biologist might experiment on fish kept in aquariums, forgetting that fish sharing a single aquarium are not independent: their conditions may be affected by one another, as well as the tested treatment.[2] If these experiments are meant to reveal trends in rats or fish in general, their results will be misleading.

---

*I just wanted an excuse to use the word *sphygmomanometer*.

You can think of pseudoreplication as collecting data that answers the wrong question. Animal behaviorists frequently try to understand bird calls, for example, by playing different calls to birds and evaluating their reactions. Bird calls can vary between geographical regions, just like human accents, and these dialects can be compared. Prior to the 1990s, a common procedure for these experiments was to record one representative bird song from each dialect and then play these songs to 10 or 20 birds and record their reactions.[3] The more birds that were observed, the larger the sample size.

But the research question was about the different song dialects, not individual songs. No matter how "representative" any given song may have been, playing it to more birds couldn't provide evidence that Dialect A was more attractive to male yellow-bellied sapsuckers than Dialect B was; it was only evidence for *that specific song or recording*. A proper answer to the research question would have required many samples of songs from both dialects.

Pseudoreplication can also be caused by taking separate measurements of the same subject over time (*autocorrelation*), like in my blood pressure experiment. Blood pressure measurements of the same patient from day to day are autocorrelated, as are revenue figures for a corporation from year to year. The mathematical structure of these autocorrelations can be complicated and vary from patient to patient or from business to business. The unwitting scientist who treats this data as though each measurement is independent of the others will obtain pseudoreplicated—and hence misleading—results.

Accounting for Pseudoreplication

Careful experimental design can break the dependence between measurements. An agricultural field experiment might compare growth rates of different strains of a crop in each field. But if soil or irrigation quality varies from field to field, you won't be able to separate variations due to crop variety from variations in soil conditions, no matter how many plants you measure in each field. A better design would be to divide each field into small blocks and randomly assign a crop variety to each block. With a large enough selection of blocks, soil variations can't systematically benefit one crop more than the others.

Alternatively, if you can't alter your experimental design, statistical analysis can help account for pseudoreplication. Statistical techniques do not magically eliminate dependence between measurements or allow you to obtain good results with poor experimental design. They merely provide ways to quantify dependence so you can correctly interpret your data. (This means they usually give wider confidence intervals and larger p values than the naive analysis.) Here are some options:[4]

- **Average the dependent data points.** For example, average all the blood pressure measurements taken from a single person and treat the average as a single data point. This isn't perfect: if you measured some patients more frequently than others, this fact won't be reflected in the averaged number. To make your results reflect the level of certainty in your measurements, which increases as you take more, you'd perform a weighted analysis, weighting the better-measured patients more strongly.

- **Analyze each dependent data point separately.** Instead of combining all the patient's blood pressure measurements, analyze every patient's blood pressure from, say, just day five, ignoring all other data points. But be careful: if you repeat this for each day of measurements, you'll have problems with multiple comparisons, which I will discuss in the next chapter.

- **Correct for the dependence by adjusting your p values and confidence intervals.** Many procedures exist to estimate the size of the dependence between data points and account for it, including clustered standard errors, repeated measures tests, and hierarchical models.[5,6,7]

Batch Biology

New technology has led to an explosion of data in biology. Inexpensive labs-on-a-chip called microarrays allow biologists to track the activities of thousands of proteins or genes simultaneously. Microarrays contain thousands of *probes*, which chemically bind to different proteins or genes; fluorescent dyes allow a scanner to detect the quantity of material bound to each probe. Cancer research in particular has benefited from these new technologies: researchers can track the expression of thousands of genes in both cancerous and healthy cells, which might lead to new targeted cancer treatments that leave healthy tissue unharmed.

Microarrays are usually processed in batches on machines that detect the fluorescent dyes. In a large study, different microarrays may be processed by different laboratories using different equipment. A naive experimental setup might be to collect a dozen cancerous samples and a dozen healthy samples, inject them into microarrays, and then run all the cancerous samples through the processing machine on Tuesday and the healthy samples on Wednesday.

You can probably see where this is going. Microarray results vary strongly between processing batches: machine calibrations might change, differences in laboratory temperature can affect chemical reactions, and different bottles of chemical reagents might be used while processing the microarrays. Sometimes the largest source of variation in an experiment's data is simply what day the microarrays were processed. Worse, these problems do not affect the entire microarray in the same way— in fact, correlations between the activity of pairs of genes can entirely *reverse* when processed in a different batch.[8] As a result, additional samples don't necessarily add data points to a biological experiment. If the new samples are processed in the same batch as the old, they just measure systematic error introduced by the equipment—not anything about cancerous cells in general.

Again, careful experimental design can mitigate this problem. If two different biological groups are being tested, you can split each group evenly between batches so systematic differences do not affect the groups in different ways. Also, be sure to record how each batch was processed, how each sample was stored, and what chemical reagents were used during processing; make this information available to the statisticians analyzing the data so they use it to detect problems.

For example, a statistician could perform principal components analysis on the data to determine whether different batches gave wildly different results. Principal components analysis determines which combinations of variables in the data account for the most variation in the results. If it indicates that the batch number is highly influential, the data can be analyzed taking batch number into account as a confounding variable.

Synchronized Pseudoreplication

Pseudoreplication can occur through less obvious routes. Consider one example in an article reviewing the prevalence of pseudoreplication in the ecological literature.[9] Suppose you want to see whether chemicals in the growing shoots of grasses are responsible for the start of the reproductive season in cute furry rodents: your hypothesis is that when the grasses sprout in springtime, the rodents eat them and begin their mating season. To test this, you try putting some animals in a lab, feed half of them ordinary food and the other half food mixed with the grasses, and wait to see when their reproductive cycles start.

But wait: you vaguely recall having read a paper suggesting that the reproductive cycles of mammals living in groups can synchronize—something about their pheromones. So maybe the animals in each group aren't actually independent of each other. After all, they're all in the same lab, exposed to the same pheromones. As soon as one goes into estrus, its pheromones could cause others to follow, no matter what they've been eating. Your sample size will be effectively one.

The research you're thinking of is a famous paper from the early 1970s, published in *Nature* by Martha McClintock, which suggested that women's menstrual cycles can synchronize if they live in close contact.[10] Other studies found similar results in golden hamsters, Norway rats, and chimpanzees. These results seem to suggest that synchronization could cause pseudoreplication in your study. Great. So does this mean you'll have to build pheromone-proof cages to keep your rodents isolated from each other?

Not quite. You might wonder how you prove that menstrual or estrous cycles synchronize. Well, as it turns out, you can't. The studies "proving" synchronization in various animals were themselves pseudoreplicated in an insidious way.

McClintock's study of human menstrual cycles went something like this:

1. Find groups of women who live together in close contact—for instance, college students in dormitories.

2. Every month or so, ask each woman when her last menstrual period began and to list the other women with whom she spent the most time.

3. Use these lists to split the women into groups that tend to spend time together.

4. For each group of women, see how far the average woman's period start date deviates from the average.

Small deviations would mean the women's cycles were aligned, all starting at around the same time. Then the researchers tested whether the deviations decreased over time, which would indicate that the women were synchronizing. To do this, they checked the mean deviation at five different points throughout the study, testing whether the deviation decreased more than could be expected by chance.

Unfortunately, the statistical test they used assumed that if there was no synchronization, the deviations would *randomly* increase and decrease from one period to another. But imagine two women in the study who start with aligned cycles. One has an average gap of 28 days between periods and the other a gap of roughly 30 days. Their cycles will diverge *consistently* over the course of the study, starting two days apart, then four days, and so on, with only a bit of random variation because periods are not perfectly timed. Similarly, two women can start the study *not* aligned but gradually align.

For comparison, if you've ever been stuck in traffic, you've probably seen how two turn signals blinking at different rates will gradually synchronize and then go out of phase again. If you're stuck at the intersection long enough, you'll see this happen multiple times. But to the best of my knowledge, there are no turn signal pheromones.

So we would actually *expect* two unaligned menstrual cycles to fall into alignment, at least temporarily. The researchers failed to account for this effect in their statistical tests.

They also made an error calculating synchronization at the beginning of the study: if one woman's period started four days before the study began and another's started four days *after*, the difference is only eight days. But periods before the beginning of the study were not counted, so the recorded difference was between the fourth day and the first woman's next period, as much as three weeks later.

These two errors combined meant that the scientists were able to obtain statistically significant results even when there was no synchronization effect outside what would occur without pheromones.[11,12]

The additional data points the researchers took as they followed subjects through more menstrual cycles did not provide evidence of synchronization at all. It was merely more statistical evidence of the synchronization that would've happened by chance, regardless of pheromones. The statistical test addressed a different question than the scientists intended to ask.

Similar problems exist with studies claiming that small furry mammals or chimpanzees synchronize their estrous cycles. Subsequent research using corrected statistical methods has failed to find any evidence of estrous or menstrual synchronization (though this is controversial).[13] We only thought our rodent experiment could have pseudoreplication because we believed a pseudoreplicated study.

Don't scoff at your friends if they complain about synchronized periods, though. If the average cycle lasts 28 days, then two average women can have periods which start at most 14 days apart. (If your period starts 20 days after your friend's, it's only eight days before her next period.) That's the maximum, so the average will be seven days, and since periods can last for five to seven days, they will frequently overlap even as cycles converge and diverge over time.

TIPS
- Ensure that your statistical analysis really answers your research question. Additional measurements that are highly dependent on previous data do not prove that your results generalize to a wider population—they merely increase your certainty about the specific sample you studied.

- Use statistical methods such as hierarchical models and clustered standard errors to account for a strong dependence between your measurements.

- Design experiments to eliminate hidden sources of correlation between variables. If that's not possible, record confounding factors so they can be adjusted for statistically. But if you don't consider the dependence from the beginning, you may find there is no way to save your data.

4

THE P VALUE AND THE BASE RATE FALLACY

You've seen that *p* values are hard to interpret. Getting a statistically insignificant result doesn't mean there's no difference between two groups. But what about getting a significant result?

Suppose I'm testing 100 potential cancer medications. Only 10 of these drugs actually work, but I don't know which; I must perform experiments to find them. In these experiments, I'll look for $p < 0.05$ gains over a placebo, demonstrating that the drug has a significant benefit.

Figure 4-1 illustrates the situation. Each square in the grid represents one drug. In reality, only the 10 drugs in the top row work. Because most trials can't perfectly detect every good medication, I'll assume my tests have a statistical power of 0.8, though you know that most studies have much lower power. So of the 10 good drugs, I'll correctly detect around 8 of them, shown in darker gray.

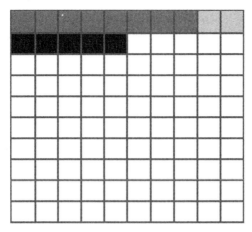

Figure 4-1: Each square represents one candidate drug. The first row of the grid represents drugs that definitely work, but I obtained statistically significant results for only the eight darker-gray drugs. The black cells are false positives.

Because my p value threshold is 0.05, I have a 5% chance of falsely concluding that an ineffective drug works. Since 90 of my tested drugs are ineffective, this means I'll conclude that about 5 of them have significant effects. These are shown in black.

I perform my experiments and conclude there are 13 "working" drugs: 8 good drugs and 5 false positives. The chance of any given "working" drug being truly effective is therefore 8 in 13—just 62%! In statistical terms, my *false discovery rate*—the fraction of statistically significant results that are really false positives—is 38%.

Because the *base rate* of effective cancer drugs is so low (only 10%), I have many opportunities for false positives. Take this to the extreme: if I had the bad fortune of getting a truckload of completely ineffective medicines, for a base rate of 0%, then I have *no* chance of getting a true significant result. Nevertheless, I'll get a $p < 0.05$ result for 5% of the drugs in the truck.

The Base Rate Fallacy

You often see news articles quoting low p values as a sign that error is unlikely: "There's only a 1 in 10,000 chance this result arose as a statistical fluke, because $p = 0.0001$." No! This can't be true. In the cancer medication example, a $p < 0.05$ threshold resulted in a 38% chance that any given statistically significant result was a fluke. This misinterpretation is called the *base rate fallacy*.

Remember how p values are defined: the p value is the probability, under the assumption that there is no true effect or no true difference, of collecting data that shows a difference equal to or more extreme than what you actually observed.

A p value is calculated under the assumption that the medication *does not work*. It tells me the probability of obtaining my data or data more extreme than it. It does *not* tell me the chance my medication is effective. A small p value is stronger evidence, but to calculate the probability that the medication is effective, you'd need to factor in the base rate.

When news came from the Large Hadron Collider that physicists had discovered evidence for the Higgs boson, a long-theorized fundamental particle, every article tried to quote a probability: "There's only a 1 in 1.74 million chance that this result is a fluke," or something along those lines. But every news source quoted a different number. Not only did they ignore the base rate and misinterpret the p value, but they couldn't calculate it correctly either.

So when someone cites a low p value to say their study is probably right, remember that the probability of error is actually almost certainly higher. In areas where most tested hypotheses are false, such as early drug trials (most early drugs don't make it through trials), it's likely that *most* statistically significant results with $p < 0.05$ are actually flukes.

A Quick Quiz

A 2002 study found that an overwhelming majority of statistics students—and instructors—failed a simple quiz about p values.[1] Try the quiz (slightly adapted for this book) for yourself to see how well you understand what p really means.

Suppose you're testing two medications, Fixitol and Solvix. You have two treatment groups, one that takes Fixitol and one that takes Solvix, and you measure their performance on some standard task (a fitness test, for instance) afterward. You compare the mean score of each group using a simple significance test, and you obtain $p = 0.01$, indicating there is a statistically significant difference between means.

Based on this, decide whether each of the following statements is true or false:

1. You have absolutely disproved the null hypothesis ("There is no difference between means").

2. There is a 1% probability that the null hypothesis is true.

3. You have absolutely proved the alternative hypothesis ("There *is* a difference between means").

4. You can deduce the probability that the alternative hypothesis is true.

5. You know, if you decide to reject the null hypothesis, the probability that you are making the wrong decision.

6. You have a reliable experimental finding, in the sense that if your experiment were repeated many times, you would obtain a significant result in 99% of trials.

You can find the answers in the footnote.*

The Base Rate Fallacy in Medical Testing

There has been some controversy over the use of mammograms to screen for breast cancer. Some argue that the dangers of false positive results—which result in unnecessary biopsies, surgery, and chemotherapy—outweigh the benefits of early cancer detection; physicians groups and regulatory agencies, such as the United States Preventive Services Task Force, have recently stopped recommending routine mammograms for women younger than 50. This is a statistical question, and the first step to answering it to ask a simpler question: if your mammogram turns up signs of cancer, what is the probability you actually have breast cancer? If this probability is too low, most positive results will be false, and a great deal of time and effort will be wasted for no benefit.

Suppose 0.8% of women who get mammograms have breast cancer. In 90% of women with breast cancer, the mammogram will correctly detect it. (That's the statistical power of the test. This is an estimate, since it's hard to tell how many cancers we miss if we don't know they're there.) However, among women with no breast cancer at all, about 7% will still get a positive reading on the mammogram. (This is equivalent to having a

_____

*I hope you've concluded that *every* statement is false. The first five statements ignore the base rate, while the last question is asking about the *power* of the experiment, not its *p* value.

$p < 0.07$ significance threshold.) If you get a positive mammogram result, what are the chances you have breast cancer?

Ignoring the chance that you, the reader, are male,* the answer is 9%.

How did I calculate this? Imagine 1,000 randomly selected women chose to get mammograms. On average, 0.8% of screened women have breast cancer, so about 8 women in our study will. The mammogram correctly detects 90% of breast cancer cases, so about 7 of the 8 will have their cancer discovered. However, there are 992 women without breast cancer, and 7% will get a false positive reading on their mammograms. This means about 70 women will be incorrectly told they have cancer.

In total, we have 77 women with positive mammograms, 7 of whom actually have breast cancer. Only 9% of women with positive mammograms have breast cancer.

Even doctors get this wrong. If you ask them, two-thirds will erroneously conclude that a $p < 0.05$ result implies a 95% chance that the result is true.[2] But as you can see in these examples, the likelihood that a positive mammogram means cancer depends on the proportion of women who actually *have* cancer. And we are very fortunate that only a small proportion of women have breast cancer at any given time.

How to Lie with Smoking Statistics

Renowned experts in statistics fall prey to the base rate fallacy, too. One high-profile example involves journalist Darrell Huff, author of the popular 1954 book *How to Lie with Statistics*.

Although *How to Lie with Statistics* didn't focus on statistics in the academic sense of the term—it was perhaps better titled *How to Lie with Charts, Plots, and Misleading Numbers*—the book was still widely adopted in college courses and read by a public eager to outsmart marketers and politicians, turning Huff into a recognized expert in statistics. So when the US Surgeon General's famous report *Smoking and Health* came out in 1964, saying that tobacco smoking causes lung cancer, tobacco companies turned to Huff to provide their public rebuttal.†

Attempting to capitalize on Huff's respected status, the tobacco industry commissioned him to testify before Congress

---

*Being male doesn't actually exclude you from getting breast cancer, but it's far less likely.
†The account that follows is based on letters and reports from the Legacy Tobacco Documents Library, an online collection of tobacco industry documents created as a result of the Tobacco Master Settlement Agreement.

and then to write a book, tentatively titled *How to Lie with Smoking Statistics*, covering the many statistical and logical errors alleged to be found in the surgeon general's report. Huff completed a manuscript, for which he was paid more than $9,000 (roughly $60,000 in 2014 dollars) by tobacco companies and which was positively reviewed by University of Chicago statistician (and paid tobacco industry consultant) K.A. Brownlee. Although it was never published, it's likely that Huff's friendly, accessible style would have made a strong impression on the public, providing talking points for watercooler arguments.

In his Chapter 7, he discusses what he calls *overprecise figures*—those presented without a confidence interval or any indication of uncertainty. For example, the surgeon general's report mentions a "mortality ratio of 1.20," which is "statistically significant at the 5 percent level." This, presumably, meant that the ratio was significantly different from 1.0, with $p < 0.05$. Huff agrees that expressing the result as a mortality ratio is perfectly proper but states:

> It does have an unfortunate result: it makes it appear that we now know the actual mortality ratio of two kinds of groups right down to a decimal place. The reader must bring to his interpretation of this figure a knowledge that what looks like a rather exact figure is only an approximation. From the accompanying statement of significance ("5 percent level") we discover that all that is actually known is that the odds are 19 to one that the second group truly does have a higher death rate than the first. The actual increase from one group to the other may be much less than the 20 percent indicated, or it may be more.

For the first half of this quote, I wanted to cheer Huff on: yes, statistically significant *doesn't* mean that we know the precise figure to two decimal places. (A confidence interval would have been a much more appropriate way to express this figure.) But then Huff claims that the significance level gives 19-to-1 odds that the death rate really is different. That is, he interprets the p value as the probability that the results are a fluke.

Not even Huff is safe from the base rate fallacy! We don't know the odds that "the second group truly does have a higher death rate than the first." All we know is that if the true mortality ratio were 1, we would observe a mortality ratio larger than 1.20 in only 1 in 20 experiments.

Huff's complaint about overprecise figures is, in fact, impossibly precise. Notably, K.A. Brownlee read this comment—and several similar remarks Huff makes throughout the manuscript—without complaint. Instead, he noted that in one case Huff incorrectly quotes the odds as 20 to 1 rather than 19 to 1. He did not seem to notice the far more fundamental base rate fallacy lurking.

Taking Up Arms Against the Base Rate Fallacy

You don't have to be performing advanced cancer research or early cancer screenings to run into the base rate fallacy. What if you're doing social research? Say you'd like to survey Americans to find out how often they use guns in self-defense. Gun control arguments, after all, center on the right to self-defense, so it's important to determine whether guns are commonly used for defense and whether that use outweighs the downsides, such as homicides.

One way to gather this data would be through a survey. You could ask a representative sample of Americans whether they own guns and, if so, whether they've used the guns to defend their homes in burglaries or themselves from being mugged. You could compare these numbers to law enforcement statistics of gun use in homicides and make an informed decision about whether the benefits of gun control outweigh the drawbacks.

Such surveys have been done, with interesting results. One 1992 telephone survey estimated that American civilians used guns in self-defense up to 2.5 million times that year. Roughly 34% of these cases were burglaries, meaning 845,000 burglaries were stymied by gun owners. But in 1992, there were only 1.3 million burglaries committed while someone was at home. Two-thirds of these occurred while the homeowners were asleep and were discovered only after the burglar had left. That leaves 430,000 burglaries involving homeowners who were at home and awake to confront the burglar, 845,000 of which, we are led to believe, were stopped by gun-toting residents.[3]

Whoops.

One explanation could be that burglaries are dramatically underreported. The total number of burglaries came from the National Crime Victimization Survey (NCVS), which asked tens of thousands of Americans in detailed interviews about their experiences with crime. Perhaps respondents who fended off a burglar with their firearms didn't report the crime—after all, nothing was stolen, and the burglar fled. But a massive underreporting of burglaries would be needed to explain the discrepancy. Fully two-thirds of burglaries committed against awake homeowners would need to have gone unreported.

A more likely answer is that the survey overestimated the use of guns in self-defense. How? In the same way mammograms overestimate the incidence of breast cancer: there are far more opportunities for false positives than false negatives. If 99.9% of people did not use a gun in self-defense in the past year but 2% of those people answered "yes" for whatever reason (to amuse themselves or because they misremembered an incident from long ago as happening in the past year), the true rate of 0.1% will appear to be nearly 2.1%, inflated by a factor of 21.

What about false negatives? Could this effect be balanced by people who said "no" even though they gunned down a mugger just last week? A respondent may have been carrying the firearm illegally or unwilling to admit using it to a stranger on the phone. But even then, if few people genuinely use a gun in self-defense, then there are few opportunities for false negatives. Even if half of gun users don't admit to it on the phone survey, they're vastly outnumbered by the tiny fraction of nonusers who lie or misremember, and the survey will give a result 20 times too large.

Since the false positive rate is the overwhelming error factor here, that's what criminologists focus on reducing. A good way to do so is by conducting extremely detailed surveys. The NCVS, run by the Department of Justice, uses detailed sit-down interviews where respondents are asked for details about crimes and their use of guns in self-defense. Only respondents who report being victimized are asked about how they defended themselves, and so people who may be inclined to lie about or misremember self-defense get the opportunity only if they also lie about or misremember being a victim. The NCVS also tries to detect misremembered dates (a common problem) by interviewing the same respondents periodically. If the respondent reports being the victim of a crime within the last six months, but six months ago they reported the same crime a few months prior, the interviewer can remind them of the discrepancy.

The 1992 NCVS estimated a much lower number than the phone survey—something like 65,000 incidents per year, not millions.[4] This figure includes not only defense against burglaries but also robberies, rapes, assaults, and car thefts. Even so, it is nearly 40 times smaller than the estimate provided by the telephone survey.

Admittedly, people may have been nervous to admit illegal gun use to a federal government agency; the authors of the original phone survey claimed that most defensive gun use involves illegal gun possession.[5] (This raises another research question: why are so many victims illegally carrying firearms?) This biases the NCVS survey results downward. Perhaps the truth is somewhere in between.

Unfortunately, the inflated phone survey figure is still often cited by gun rights groups, misinforming the public debate on gun safety. Meanwhile, the NCVS results hold steady at far lower numbers. The gun control debate is far more complicated than a single statistic, of course, but informed debate can begin only with accurate data.

If At First You Don't Succeed, Try, Try Again

The base rate fallacy shows that statistically significant results are false positives much more often than the $p < 0.05$ criterion for significance might suggest. The fallacy's impact is magnified in modern research, which usually doesn't make just one significance test. More often, studies compare a variety of factors, seeking those with the most important effects.

For example, imagine testing whether jelly beans cause acne by testing the effect of every single jelly bean color on acne, as illustrated in Figure 4-2.

Figure 4-2: Cartoon from xkcd, by Randall Munroe
(http://xkcd.com/882/)

As the comic shows, making multiple comparisons means multiple chances for a false positive. The more tests I perform, the greater the chance that at least one of them will produce a false positive. For example, if I test 20 jelly bean flavors that do not cause acne at all and look for a correlation at $p < 0.05$ significance, I have a 64% chance of getting at least one false positive result. If I test 45 flavors, the chance of at least one false positive is as high as 90%. If I instead use confidence intervals to look for a correlation that is nonzero, the same problem will occur.

NOTE   *The math behind these numbers is fairly straightforward. Suppose we have* n *independent hypotheses to test, none of which is true. We set our significance criterion at* p < 0.05. *The probability of obtaining at least one false positive among the* n *tests is as follows:*

$$P(\text{false positive}) = 1 - (1 - 0.05)^n$$

For n = 100, *the false positive probability increases to 99%.*

Multiple comparisons aren't always as obvious as testing 20 jelly bean colors. Track the symptoms of patients for a dozen weeks and test for significant benefits during any of those weeks: bam, that's 12 comparisons. And if you're checking for the occurrence of 23 different potential dangerous side effects? Alas! You have sinned.

If you send out a 10-page survey asking about nuclear power plant proximity, milk consumption, age, number of male cousins, favorite pizza topping, current sock color, and a few dozen other factors for good measure, you'll probably find that at least one of those things is correlated with cancer.

Particle physicists call this the *look-elsewhere effect*. An experiment like the Large Hadron Collider's search for the Higgs boson involves searching particle collision data, looking for small anomalies that indicate the existence of a new particle. To compute the statistical significance of an anomaly at an energy of 5 gigaelectronvolts,* for example, physicists ask this: "How likely is it to see an anomaly this size or larger at 5 gigaelectronvolts by chance?" But they could have looked elsewhere—they are searching for anomalies across a large swath of energies, any one of which could have produced a

---

*Physicists have the best unit names. Gigaelectronvolts, jiffies, inverse femtobarns—my only regret as a physicist who switched to statistics is that I no longer have excuses to use these terms.

false positive. Physicists have developed complicated procedures to account for this and correctly limit the false positive rate.[6]

If we want to make many comparisons at once but control the *overall* false positive rate, the p value should be calculated under the assumption that *none* of the differences is real. If we test 20 different jelly beans, we would not be surprised if one out of the 20 "causes" acne. But when we calculate the p value for a specific flavor, as though each comparison stands on its own, we are calculating the probability that *this specific* group would be lucky—an unlikely event—not any 1 out of the 20. And so the anomalies we detect appear much more significant than they are.[7]

A survey of medical trials in the 1980s found that the average trial made 30 therapeutic comparisons. In more than half the trials, the researchers had made so many comparisons that a false positive was highly likely, casting the statistically significant results they did report into doubt. They may have found a statistically significant effect, but it could just have easily been a false positive.[8] The situation is similar in psychology and other heavily statistical fields.

There are techniques to correct for multiple comparisons. For example, the Bonferroni correction method allows you to calculate p values as you normally would but says that if you make n comparisons in the trial, your criterion for significance should be $p < 0.05/n$. This lowers the chances of a false positive to what you'd see from making only one comparison at $p < 0.05$. However, as you can imagine, this reduces statistical power, since you're demanding much stronger correlations before you conclude they're statistically significant. In some fields, power has decreased systematically in recent decades because of increased awareness of the multiple comparisons problem.

In addition to these practical problems, some researchers object to the Bonferroni correction on philosophical grounds. The Bonferroni procedure implicitly assumes that *every* null hypothesis tested in multiple comparisons is true. But it's almost never the case that the difference between two populations is exactly zero or that the effect of some drug is exactly identical to a placebo. So why assume the null hypothesis is true in the first place?

If this objection sounds familiar, it's because you've heard it before—as an argument against null hypothesis significance testing *in general*, not just the Bonferroni correction. Accurate estimates of the *size* of differences are much more interesting than checking only whether each effect could be zero. That's all the more reason to use confidence intervals and effect size estimates instead of significance testing.

Red Herrings in Brain Imaging

Neuroscientists do massive numbers of comparisons when performing functional MRI (fMRI) studies, where a three-dimensional image of the brain is taken before and after the subject performs some task. The images show blood flow in the brain, revealing which parts of the brain are most active when a person performs different tasks.

How exactly do you decide which regions of the brain are active? A simple method is to divide the brain image into small cubes called voxels. A voxel in the "before" image is compared to the voxel in the "after" image, and if the difference in blood flow is significant, you conclude that part of the brain was involved in the task. Trouble is, there are tens of thousands of voxels to compare and therefore many opportunities for false positives.

One study, for instance, tested the effects of an "open-ended mentalizing task" on participants. Subjects were shown "a series of photographs depicting human individuals in social situations with a specified emotional valence" and asked to "determine what emotion the individual in the photo must have been experiencing." You can imagine how various emotional and logical centers of the brain would light up during this test.

The data was analyzed, and certain brain regions were found to change activity during the task. Comparison of images made before and after the "mentalizing task" showed a $p = 0.001$ difference in an $81mm^3$ cluster in the brain.

The study participants? Not college undergraduates paid $10 for their time, as is usual. No, the test subject was a 3.8-pound Atlantic salmon, which "was not alive at the time of scanning."[*]

Neuroscientists often attempt to limit this problem by requiring clusters of 10 or more significant voxels with a stringent threshold of $p < 0.005$, but in a brain scan with tens of thousands of voxels, a false positive is still virtually guaranteed. Techniques like the Bonferroni correction, which control the

---

[*] "Foam padding was placed within the head coil as a method of limiting salmon movement during the scan, but proved to be largely unnecessary as subject motion was exceptionally low."

rate of false positives even when thousands of statistical tests are made, are now common in the neuroscience literature. Few papers make errors as serious as the ones demonstrated in the dead salmon experiment. Unfortunately, almost every paper tackles the problem differently. One review of 241 fMRI studies found that they used 207 unique combinations of statistical methods, data collection strategies, and multiple comparison corrections, giving researchers great flexibility to achieve statistically significant results.[9]

Controlling the False Discovery Rate

As I mentioned earlier, one drawback of the Bonferroni correction is that it greatly decreases the statistical power of your experiments, making it more likely that you'll miss true effects. More sophisticated procedures than Bonferroni correction exist, ones with less of an impact on statistical power, but even these are not magic bullets. Worse, they don't spare you from the base rate fallacy. You can still be misled by your p threshold and falsely claim there's "only a 5% chance I'm wrong." Procedures like the Bonferroni correction only help you eliminate some false positives.

Scientists are more interested in limiting the *false discovery rate*: the fraction of statistically significant results that are false positives. In the cancer medication example that started this chapter, my false discovery rate was 38%, since fully one-third of my statistically significant results were flukes. Of course, the only reason you knew how many of the medications *actually* worked was because I told you the number ahead of time. In general, you don't know how many of your tested hypotheses are true; you can compute the false discovery rate only by guessing. But ideally, you'd find it out from the data.

In 1995, Yoav Benjamini and Yosef Hochberg devised an exceptionally simple procedure that tells you which p values to consider statistically significant. I've been saving you from mathematical details so far, but to illustrate just how simple the procedure is, here it is:

1. Perform your statistical tests and get the p value for each. Make a list and sort it in ascending order.

2. Choose a false-discovery rate and call it q. Call the number of statistical tests m.

3. Find the largest p value such that $p \leq iq/m$, where i is the p value's place in the sorted list.

4. Call that p value and all smaller than it statistically significant.

You're done! The procedure guarantees that out of all statistically significant results, on average no more than q percent will be false positives.[10] I hope the method makes intuitive sense: the p cutoff becomes more conservative if you're looking for a smaller false-discovery rate (smaller q) or if you're making more comparisons (higher m).

The Benjamini–Hochberg procedure is fast and effective, and it has been widely adopted by statisticians and scientists. It's particularly appropriate when testing hundreds of hypotheses that are expected to be mostly false, such as associating genes with diseases. (The vast majority of genes have nothing to do with a particular disease.) The procedure usually provides better statistical power than the Bonferroni correction, and the false discovery rate is easier to interpret than the false positive rate.

TIPS
- Remember, $p < 0.05$ isn't the same as a 5% chance your result is false.

- If you are testing multiple hypotheses or looking for correlations between many variables, use a procedure such as Bonferroni or Benjamini–Hochberg (or one of their various derivatives and adaptations) to control for the excess of false positives.

- If your field routinely performs multiple tests, such as in neuroimaging, learn the best practices and techniques specifically developed to handle your data.

- Learn to use prior estimates of the base rate to calculate the probability that a given result is a false positive (as in the mammogram example).

5

BAD JUDGES OF SIGNIFICANCE

Using too many statistical significance tests is a good way to get misleading results, but it's also possible to claim significance for a difference you haven't explicitly tested. Misleading error bars could convince you that a test is unnecessary, or a difference in the statistical significance of two treatments might convince you there's a statistically significant difference between them. Let's start with the latter.

Insignificant Differences in Significance

"We compared treatments A and B with a placebo. Treatment A showed a significant benefit over placebo, while treatment B had no statistically significant benefit. Therefore, treatment A is better than treatment B."

We hear this all the time. It's an easy way of comparing medications, surgical interventions, therapies, and experimental results. It's straightforward. It seems to make sense.

However, a difference in significance does not always make a significant difference.[1]

One reason is the arbitrary nature of the $p < 0.05$ cutoff. We could get two very similar results, with $p = 0.04$ and $p = 0.06$, and mistakenly say they're clearly different from each other simply because they fall on opposite sides of the cutoff. The second reason is that p values are not measures of effect size, so similar p values do not always mean similar effects. Two results with identical statistical significance can nonetheless contradict each other.

Instead, think about statistical power. If we compare our new experimental drugs Fixitol and Solvix to a placebo but we don't have enough test subjects to give us good statistical power, then we may fail to notice their benefits. If they have identical effects but we have only 50% power, then there's a good chance we'll say Fixitol has significant benefits and Solvix does not. Run the trial again, and it's just as likely that Solvix will appear beneficial and Fixitol will not.

It's fairly easy to work out the math. Assume both drugs have identical nonzero effects compared to the placebo, and our experiments have statistical power B. This means the probability that we will detect each group's difference from control is B, so the probability that we will detect Fixitol's effect but *not* Solvix's is $B(1 - B)$. The same goes for detecting Solvix's effect but not Fixitol's. Add the probabilities up, and we find that the probability of concluding that one drug has a significant effect and the other does not is $2B(1 - B)$. The result is plotted in Figure 5-1.

Instead of independently comparing each drug to the placebo, we should compare them against each other. We can test the hypothesis that they are equally effective, or we can construct a confidence interval for the extra benefit of Fixitol over Solvix. If the interval includes zero, then they could be equally effective; if it doesn't, then one medication is a clear winner. This doesn't improve our statistical power, but it does prevent the false conclusion that the drugs are different. Our tendency to look for a difference in significance should be replaced by a check for the significance of the difference.

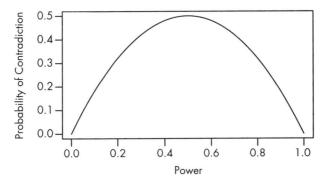

Figure 5-1: A plot of $2B(1-B)$, the probability that one drug will show a significant result and the other an insignificant result despite both drugs having identical effects. When the power is very low, both drugs give insignificant results; when the power is very high, both drugs give significant results.

This subtle distinction is important to keep in mind, for example, when interpreting the results of *replication studies*, in which researchers attempt to reproduce the results of previous studies. Some replication studies frame their negative results in terms of significance: "The original paper obtained a significant result, but this more careful study did not." But even if the replication experiment was designed to have sufficient statistical power to detect the effect reported in the initial study, there was probably truth inflation—the initial study probably overstated the effect. Since a larger sample is required to detect a smaller effect, the true power of the replication experiment may be lower than intended, and it's perfectly possible to obtain a statistically insignificant result that is nevertheless consistent with the earlier research.

As another example, in 2007 the No. 7 Protect & Perfect Beauty Serum became a best seller for Boots, the UK pharmacy chain, after the BBC reported on a clinical trial that supposedly proved its effectiveness in reducing skin wrinkles. According to the trial, published by the *British Journal of Dermatology*, the serum reduced the number of wrinkles in 43% of test subjects, a statistically significant benefit, whereas the control treatment (the same serum without the active ingredient) benefited only 22% of subjects, a statistically insignificant improvement. The implication, touted in advertising, was that the serum was scientifically proven to be your best choice for wrinkle control—even though the authors had to admit in their paper that the difference between the groups was not statistically significant.[2]

This misuse of statistics is not limited to corporate marketing departments, unfortunately. Neuroscientists, for instance, use the incorrect method for comparing groups about half the time.[3] You might also remember news about a 2006 study suggesting that men with multiple older brothers are more likely to be homosexual.[4] How did they reach this conclusion? The authors explained their results by noting that when they ran an analysis of the effect of various factors on homosexuality, only the number of older brothers had a statistically significant effect. The number of older sisters or of nonbiological older brothers (that is, adopted brothers or stepbrothers) had no statistically significant effect. But as we've seen, this doesn't guarantee there's a significant difference *between* these different effect groups. In fact, a closer look at the data suggests there was no statistically significant difference between the effect of having older brothers versus older sisters. Unfortunately, not enough data was published in the paper to allow calculation of a p value for the comparison.[1]

This misinterpretation of inconclusive results contributes to the public impression that doctors can't make up their minds about what medicines and foods are good or bad for you. For example, statin drugs have become wildly popular to reduce blood cholesterol levels because high cholesterol is associated with heart disease. But this association doesn't *prove* that reducing cholesterol levels will benefit patients. A series of five large meta-analyses reviewing tens of thousands of patient records set out to answer this question: "Do statins reduce mortality in patients who have no history of cardiovascular disease?"

Three of the studies answered yes, statins *do* reduce mortality rates. The other two concluded there was not enough evidence to suggest statins are helpful.[5] Doctors, patients, and journalists reading these articles were no doubt confused, perhaps assuming the research on statins was contradictory and inconclusive. But as the confidence intervals plotted in Figure 5-2 show, all five meta-analyses gave similar estimates of the effect of statins: the relative risk estimates were all near 0.9, indicating that during the trial periods, 10% fewer patients on statin drugs died. Although two studies did have confidence intervals overlapping a relative risk of one—indicating no difference between treatment and control—their effect size estimates matched the other studies well. It would be silly to claim there was serious disagreement between studies.

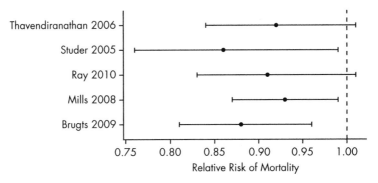

Figure 5-2: Confidence intervals for the relative risk of mortality among patients taking statin drugs, estimated by five different large meta-analyses. A relative risk of less than one indicates smaller mortality rates than among the control group. The meta-analyses are labeled by the lead author's name and year of publication.

Ogling for Significance

In the previous section, I said that if we want to compare Fixitol and Solvix, we should use a significance test to compare them directly, instead of comparing them both against placebo. Why must I do that? Why can't I just look at the two confidence intervals and judge whether they overlap? If the confidence intervals overlap, it's plausible both drugs have the same effect, so they must not be significantly different, right? Indeed, when judging whether a significant difference exists, scientists routinely eyeball it, making use of plots like Figure 5-3.

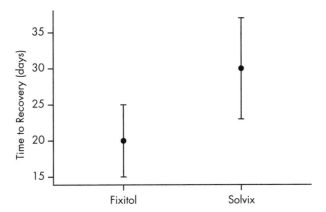

Figure 5-3: Time until recovery of patients using Fixitol or Solvix. Fixitol appears to be more effective, but the error bars overlap.

Imagine the two plotted points indicate the estimated time until recovery from some disease in two different groups of 10 patients. The width of these error bars could represent three different things.

1. Twice the standard deviation of the measurements. Calculate how far each observation is from the average, square each difference, and then average the results and take the square root. This is the standard deviation, and it measures how spread out the measurements are from their mean. Standard deviation bars stretch from one standard deviation below the mean to one standard deviation above.

2. The 95% confidence interval for the estimate.

3. Twice the standard error for the estimate, another way of measuring the margin of error. If you run numerous identical experiments and obtain an estimate of Fixitol's effectiveness from each, the standard error is the standard deviation of these estimates. The bars stretch one standard error below and one standard error above the mean. In the most common cases, a standard error bar is about half as wide as the 95% confidence interval.

It is important to notice the distinction between these. The standard deviation measures the *spread* of the individual data points. If I were measuring how long it takes for patients to get better when taking Fixitol, a high standard deviation would tell me it benefits some patients much more than others. Confidence intervals and standard errors, on the other hand, estimate how far the *average* for this sample might be from the true average—the average I would get if I could give Fixitol to every single person who ever gets the disease. Hence, it is important to know whether an error bar represents a standard deviation, confidence interval, or standard error, though papers often do not say.*

For now, let's assume Figure 5-3 shows two 95% confidence intervals. Since they overlap, many scientists would conclude there is no statistically significant difference between the groups. After all, groups one and two might not be different— the average time to recover could be 25 days in both groups, for example, and the differences appeared only because group one got lucky this time.

---

*And because standard error bars are about half as wide as the 95% confidence interval, many papers will report "standard error bars" that actually span *two* standard errors above and below the mean, making a confidence interval instead.

But does this really mean the difference isn't statistically significant? What would its p value be?

I can calculate p using a t test, the standard statistical test for telling whether the means of two groups are significantly different from each other. Plugging in the numbers for Fixitol and Solvix, I find that $p < 0.05$! There is a statistically significant difference between them, even though the confidence intervals overlap.

Unfortunately, many scientists skip the math and simply glance at plots to see whether confidence intervals overlap. Since intervals can overlap but still represent a statistically significant difference, this is actually a much more conservative test—it's always stricter than requiring $p < 0.05$.[6] And so significant differences will be missed.

Earlier, we assumed the error bars in Figure 5-3 represent confidence intervals. But what if they are standard errors or standard deviations? Could we spot a significant difference by just looking for whether the error bars overlap? As you might guess, no. For standard errors, we have the opposite problem we had with confidence interval bars: two observations might have standard errors that don't overlap, but the difference between the two is *not* statistically significant. And standard deviations do not give enough information to judge significance, whether they overlap or not.

A survey of psychologists, neuroscientists, and medical researchers found that the majority judged significance by confidence interval overlap, with many scientists confusing standard errors, standard deviations, and confidence intervals.[7] Another survey, of climate science papers, found that a majority of papers that compared two groups with error bars made this error.[8] Even introductory textbooks for experimental scientists, such as John Taylor's *An Introduction to Error Analysis*, teach students to judge by eye, hardly mentioning formal hypothesis tests at all.

There is exactly one situation when visually checking confidence intervals works, and it is when comparing the confidence interval against a fixed value, rather than another confidence interval. If you want to know whether a number is plausibly zero, you may check to see whether its confidence interval overlaps with zero. There are, of course, formal statistical procedures that generate confidence intervals that *can* be compared by eye and that even correct for multiple comparisons automatically. Unfortunately, these procedures work only in certain circumstances; Gabriel comparison intervals, for example, are easily interpreted by eye but require each group being compared to have the same standard deviation.[9] Other procedures handle more general cases, but only approximately and not in ways that can easily be plotted.[10] (The alternative, doing a separate test for each possible pair of variables and then using the Bonferroni correction for multiple comparisons, is tedious and conservative, lowering the statistical power more than alternative procedures.)

Overlapping confidence intervals do not mean two values are not significantly different. Checking confidence intervals or standard errors will mislead. It's always best to use the appropriate hypothesis test instead. Your eyeball is not a well-defined statistical procedure.

TIPS
- Compare groups directly using appropriate statistical tests, instead of simply saying, "This one was significant, and this one wasn't."
- Do not judge the significance of a difference by eye. Use a statistical test.
- Remember that if you compare many groups, you need to adjust for making multiple comparisons!

6

DOUBLE-DIPPING IN THE DATA

Earlier, we discussed truth inflation, a symptom of the overuse
of significance testing. In the quest
for significance, researchers select
only the luckiest and most exaggerated results since those are the only ones that pass
the significance filter. But that's not the only
way research gets biased toward exaggerated
results.

Statistical analyses are often *exploratory*. In exploratory data
analysis, you don't choose a hypothesis to test in advance. You
collect data and poke it to see what interesting details might
pop out, ideally leading to new hypotheses and new experiments. This process involves making numerous plots, trying a
few statistical analyses, and following any promising leads.

But aimlessly exploring data means a lot of opportunities
for false positives and truth inflation. If in your explorations
you find an interesting correlation, the standard procedure is

to collect a new dataset and test the hypothesis again. Testing an independent dataset will filter out false positives and leave any legitimate discoveries standing. (Of course, you'll need to ensure your test dataset is sufficiently powered to replicate your findings.) And so exploratory findings should be considered tentative until confirmed.

If you *don't* collect a new dataset or your new dataset is strongly related to the old one, truth inflation will come back to bite you in the butt.

Circular Analysis

Suppose I want to implant electrodes in the brain of a monkey, correlating their signals with images I'll be projecting on a screen. My goal is to understand how the brain processes visual information. The electrodes will record communication between neurons in the monkey's visual cortex, and I want to see whether different visual stimuli will result in different neuronal firing patterns. If I get statistically significant results, I might even end up in news stories about "reading monkeys' minds."

When implantable electrodes were first available, they were large and could record only a few neurons at a time. If the electrode was incorrectly placed, it might not detect any useful signal at all, so to ensure it clearly recorded neurons that had something to do with vision, it would be slowly moved as the monkey viewed a stimulus. When clear responses were seen, the electrode would be left in place and the experiment would begin. Hence, the exploratory analysis was confirmed by a full experiment.

Placing the electrode is an exploratory analysis: let's try some neurons until one of them seems to fire whenever the monkey views an image. But once the electrode is in place, we collect a new set of data and test whether, say, the neuron firing rate tells us whether the monkey is viewing a green or purple image. The new data is separate from the old, and if we simply got a lucky correlation when placing the electrode, we would fail to replicate the finding during the full experiment.

Modern electrodes are much smaller and much more sophisticated. A single implant the size of a dime contains dozens of electrodes, so we can implant the chip and afterward select the electrodes that seem to give the best signals. A modern experiment, then, might look something like this: show the monkey a variety of stimuli, and record neural responses with the electrodes. Analyze the signal from every electrode to see whether it showed any reaction above the normal background firing rate, which would indicate that it is picking up signals from a neuron we're interested in. (This analysis may be corrected for multiple comparisons to prevent high false positive rates.)

Using these results, we discard data from the electrodes that missed their targets and analyze the remaining data more extensively, testing whether the firing patterns varied with the different stimuli we presented. It's a two-stage procedure: first pick out electrodes that have a good signal and appear related to vision; then determine whether their signals vary between different stimuli. It's tempting to reuse the data we already collected, since we didn't have to move the electrodes. It's essentially a shotgun approach: use many small electrodes, and some are bound to hit the right neurons. With the bad electrodes filtered out, we can test whether the remaining electrodes appear to fire at different rates in response to the different stimuli. If they do, we've learned something about the location of vision processing in monkey brains.

Well, not quite. If I went ahead with this plan, I'd be using the same data twice. The statistical test I use to find a correlation between the neuron and visual stimulus computes p assuming no correlation—that is, it assumes the null hypothesis, that the neuron fires randomly. But after the exploratory phase, I specifically selected neurons that seem to fire *more* in reaction to the visual stimuli. In effect, I'd be testing only the lucky neurons, so I should always expect them to be associated with different visual stimuli.[1] I could do the same experiment on a dead salmon and get positive results.

This problem, *double-dipping* in the data, can cause wildly exaggerated results. And double-dipping isn't specific to neural electrodes; here's an example from fMRI testing, which aims to associate activity in specific regions of the brain with stimuli or activities. The MRI machine detects changes in blood flow to different parts of the brain, indicating which areas are working harder to process the stimulus. Because modern MRI machines provide very high-resolution images, it's important to select a region of interest in the brain in advance; otherwise, we'd be performing comparisons across tens of thousands of individual points in the brain, requiring massive multiple comparison correction and lowering the study's statistical power substantially. The region of interest may be selected on the basis of biology or previous results, but often there is no clear region to select.

Suppose, for example, we show a test subject two different stimuli: images of walruses and images of penguins. We don't know which part of the brain processes these stimuli, so we perform a simple test to see whether there is a difference between the activity caused by walruses and the activity when the subject sees no stimulus at all. We highlight regions with statistically significant results and perform a full analysis on those regions, testing whether activity patterns differ between the two stimuli.

If walruses and penguins cause equal activation in a certain region of the brain, our screening is likely to select that region for further analysis. However, our screening test also picked out regions where random variations and noise caused greater apparent activation for walruses. So our full analysis will show higher activation on average for walruses than for penguins. We will detect this nonexistent difference several times more often than the false positive rate of our test would suggest, because we are testing only the lucky regions.[2] Walruses do have a true effect, so we have not invented a spurious correlation—but we *have* inflated the size of its effect.

Of course, this is a contrived example. What if we chose the region of interest using both stimuli? Then we wouldn't mistakenly believe walruses cause greater activation than penguins. Instead, we would mistakenly overstate *both* of their effects. Ironically, using more stringent multiple comparisons corrections to select the region of interest makes the problem worse. It's the truth inflation phenomenon all over again. Regions showing average or below-average responses are not included in the final analysis, because they were insufficiently significant. Only areas with the strongest random noise make it into further analysis.

There are several ways to mitigate this problem. One is to split the dataset in half, choosing regions of interest with the first half and performing the in-depth analysis with the second. This reduces statistical power, though, so we'd have to collect more data to compensate. Alternatively, we could select regions of interest using some criterion other than response to walrus or penguin stimuli, such as prior anatomical knowledge.

These rules are often violated in the neuroimaging literature, perhaps as much as 40% of the time, causing inflated correlations and false positives.[2] Studies committing this error tend to find larger correlations between stimuli and neural activity than are plausible, given the random noise and error inherent to brain imaging.[3] Similar problems occur when geneticists collect data on thousands of genes and select subsets for analysis or when epidemiologists dredge through demographics and risk factors to find which ones are associated with disease.[4]

Regression to the Mean

Imagine tracking some quantity over time: the performance of a business, a patient's blood pressure, or anything else that varies gradually with time. Now pick a date and select all the subjects that stand out: the businesses with the highest revenues, the patients with the highest blood pressures, and so on. What happens to those subjects the next time we measure them?

Well, we've selected all the top-performing businesses and patients with chronically high blood pressure. But we've also selected businesses having an unusually lucky quarter and patients having a particularly stressful week. These lucky and unlucky subjects won't stay exceptional forever; measure them again in a few months, and they'll be back to their usual performance.

This phenomenon, called *regression to the mean*, isn't some special property of blood pressures or businesses. It's just the observation that luck doesn't last forever. On average, everyone's luck is average.

Francis Galton observed this phenomenon as early as 1869.[5] While tracing the family trees of famous and eminent people, he noticed that the descendants of famous people tended to be less famous. Their children may have inherited the great musical or intellectual genes that made their parents so famous, but they were rarely as eminent as their parents. Later investigation revealed the same behavior for heights: unusually tall parents had children who were more average, and unusually short parents had children who were usually taller.

Returning to the blood pressure example, suppose I pick out patients with high blood pressure to test an experimental drug. There are several reasons their blood pressure might be high, such as bad genes, a bad diet, a bad day, or even measurement error. Though genes and diet are fairly constant, the other factors can cause someone's measured blood pressure to vary from day to day. When I pick out patients with high blood pressure, many of them are probably just having a bad day or their blood pressure cuff was calibrated incorrectly.

And while your genes stay with you your entire life, a poorly calibrated blood pressure cuff does not. For those unlucky patients, their luck will improve soon enough, *regardless of whether I treat them or not*. My experiment is biased toward finding an effect, purely by virtue of the criterion I used to select my subjects. To correctly estimate the effect of the medication, I need to randomly split my sample into treatment and control groups. I can claim the medication works only if the treatment group has an average blood pressure improvement substantially better than the control group's.

Another example of regression to the mean is test scores. In the chapter on statistical power, I discussed how random variation is greater in smaller schools, where the luck of an individual student has a greater effect on the school's average results. This also means that if we pick out the best-performing schools—those that have a combination of good students, good teachers, and *good luck*—we can expect them to perform less well next year simply because good luck is fleeting. As is bad luck: the worst schools can expect to do better next year—which might convince administrators that their interventions worked, even though it was really only regression to the mean.

A final, famous example dates back to 1933, when the field of mathematical statistics was in its infancy. Horace Secrist, a statistics professor at Northwestern University, published *The Triumph of Mediocrity in Business*, which argued that unusually successful businesses tend to become less successful and unsuccessful businesses tend to become more successful: proof that businesses trend toward mediocrity. This was not a statistical artifact, he argued, but a result of competitive market forces. Secrist supported his argument with reams of data and numerous charts and graphs and even cited some of Galton's work in regression to the mean. Evidently, Secrist did not understand Galton's point.

Secrist's book was reviewed by Harold Hotelling, an influential mathematical statistician, for the *Journal of the American Statistical Association*. Hotelling pointed out the fallacy and noted that one could easily use the same data to prove that business trend *away* from mediocrity: instead of picking the best businesses and following their decline over time, track their progress from *before* they became the best. You will invariably find that they improve. Secrist's arguments "really prove nothing more than that the ratios in question have a tendency to wander about."[5]

Stopping Rules

Medical trials are expensive. Supplying dozens of patients with experimental medications and tracking their symptoms over the course of months takes significant resources, so many pharmaceutical companies develop *stopping rules*, which allow investigators to end a study early if it's clear the experimental drug has a substantial effect. For example, if the trial is only half complete but there's already a statistically significant difference in symptoms with the new medication, the researchers might terminate the study rather than gathering more data to reinforce the conclusion. In fact, it is considered *unethical* to withhold a medication from the control group if you already know it to be effective.

If poorly done, however, dipping into the data early can lead to false positives.

Suppose we're comparing two groups of patients, one taking our experimental new drug Fixitol and one taking a placebo. We measure the level of some protein in their bloodstreams to see whether Fixitol is working. Now suppose Fixitol causes no change whatsoever and patients in both groups have the same average protein levels. Even so, protein levels will vary slightly among individuals.

We plan to use 100 patients in each group, but we start with 10, gradually recruiting additional pairs to place in the treatment and control groups. As we go along, we do a significance test to compare the two groups and see whether there is a statistically significant difference between average protein levels. We'll stop early if we see statistical significance. We might see a result like the simulation in Figure 6-1.

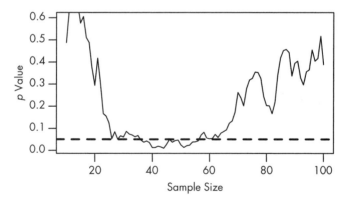

Figure 6-1: The results of a significance test taken after every pair of new patients is added to the study. There is no genuine difference between groups. The dashed line indicates the p = 0.05 significance level.

The plot shows the *p* value of the difference between groups as we collect more data, with the dashed line indicating the $p = 0.05$ level of significance. At first, there appears to be no significant difference. But as we collect more and more data, the *p* value dips below the dashed line. If we were to stop early, we'd mistakenly conclude that there is a significant difference between groups. Only as we collect even more data do we realize the difference isn't significant.

You might expect that the p value dip shouldn't happen since there's no real difference between groups. After all, taking more data shouldn't make our conclusions worse, right? It's true that if we run the trial again, we might find that the groups start out with no significant difference and stay that way as we collect more data or that they start with a huge difference and quickly regress to having none. But if we wait long enough and test after every data point, we will eventually cross *any* arbitrary line of statistical significance. We can't usually collect infinite samples, so in practice this doesn't always happen, but poorly implemented stopping rules still increase false positive rates significantly.[6]

Our intent in running the experiment is important here. Had we chosen a fixed group size in advance, the p value would be the probability of obtaining more extreme results with that particular group size. But since we allowed the group size to vary depending on the results, the p value has to be calculated taking this into account. An entire field of sequential analysis has developed to solve these problems, either by choosing a more stringent p value threshold that accounts for the multiple testing or by using different statistical tests.

Aside from false positives, trials with early stopping rules also tend to suffer disproportionately from truth inflation. Many trials that are stopped early are the result of lucky patients, not brilliant drugs. By stopping the trial, researchers have deprived themselves of the extra data needed to tell the difference. In fact, stopped medical trials exaggerate effects by an average of 29% over similar studies that are not stopped early.[7]

Of course, we don't know the Truth about any drug being studied. If we did, we wouldn't be running the study in the first place! So we can't tell whether a particular study was stopped early because of luck or because the drug really was good. But many stopped studies don't even publish their original intended sample size or the stopping rule used to justify terminating the study.[8] A trial's early stoppage is not automatic evidence that its results are biased, but it *is* suggestive.

Modern clinical trials are often required to register their statistical protocols in advance and generally preselect only a few evaluation points at which to test their evidence, rather than after every observation. Such *registered studies* suffer only a small increase in the false positive rate, which can be accounted for by carefully choosing the required significance levels and other sequential analysis techniques.[9] But most other fields do not use protocol registration, and researchers have the freedom to use whatever methods they feel appropriate. For example, in a survey of academic psychologists, more than half admitted to deciding whether to collect more data after checking whether their results were significant, usually concealing this practice in publications.[10] And given that researchers probably aren't eager to admit to questionable research practices, the true proportion is likely higher.

TIPS
- If you use your data to decide on your analysis procedure, use separate data to perform the analysis.

- If you use a significance test to pick out the luckiest (or unluckiest) people in your sample of data, don't be surprised if their luck doesn't hold in future observations.

- Carefully plan stopping rules in advance and adjust for multiple comparisons.

7

CONTINUITY ERRORS

So far in this book, I've focused on comparisons between groups. Is the placebo or the drug more effective? Do intersections that allow right turns on red kill more people than those that don't? You produce a single statistic for each group—such as an average number of traffic accidents—and see whether these statistics are significantly different between groups.

But what if you can't separate test subjects into clear groups? A study of the health impacts of obesity might measure the body mass index of each participant, along with blood pressure, blood sugar, resting heart rate, and so on. But there aren't two clear groups of patients; there's a spectrum, from underweight to obese. Say you want to spot health trends as you move from one end of this spectrum to the other.

One statistical technique to deal with such scenarios is called *regression modeling*. It estimates the *marginal* effect of each variable—the health impact of each additional pound of weight, not just the difference between groups on either side of an arbitrary cutoff. This gives much finer-grained results than a simple comparison between groups.

But scientists frequently simplify their data to avoid the need for regression analysis. The statement "Overweight people are 50% more likely to have heart disease" has far more obvious clinical implications than "Each additional unit of Metropolitan Relative Weight increases the log-odds of heart disease by 0.009." Even if it's possible to build a statistical model that captures every detail of the data, a statistician might choose a simpler analysis over a technically superior one for purely practical reasons. As you've seen, simple models can still be used incorrectly, and the process of simplifying the data introduces yet more room for error. Let's start with the simplification process; in the next chapter, I'll discuss common errors when using full regression models instead.

Needless Dichotomization

A common simplification technique is to *dichotomize* variables by splitting a continuous measurement into two separate groups. In the example study on obesity, for example, you might divide patients into "healthy" or "overweight" groups. By splitting the data, you don't need to fuss over choosing the correct regression model. You can just compare the two groups using a *t* test.

This raises the question: how do you decide where to split the data? Perhaps there's a natural cutoff or a widely accepted definition (as with obesity), but often there isn't. One common solution is to split the data along the median of the sample, which divides the data into two equal-size groups—a so-called *median split.* A downside to this approach is that different researchers studying the same phenomenon will arrive at different split points, making their results difficult to compare or aggregate in meta-analyses.

An alternative to a median split is to select the cutoff that gives you the smallest *p* value between groups. You can think of this as choosing to separate the groups so they are the "most different." As you might imagine, this approach makes false positives more likely. Searching for the cutoff with the best *p* value means effectively performing many hypothesis tests until you get the result you want. The result is the same as you saw previously with multiple comparisons: a false positive rate increased by as much as a factor of 10.[1] Your confidence intervals for the effect size will also be misleadingly narrow.

Dichotomization problems cropped up in a number of breast cancer research papers in the early 1990s studying the S-phase fraction, the fraction of cells in a tumor that are busy copying and synthesizing new DNA. Oncologists believe this fraction may predict the ultimate course of a cancer, allowing doctors to target their patients' treatments more effectively. Researchers studying the matter divided patients into two groups: those with large S-phase fractions and those with small ones.

Of course, each study chose a different cutoff between "large" and "small," picking either the median or the cutoff that gave the best *p* value. Unsurprisingly, the studies that chose the "optimal" cutoff had statistically significant results. But when these were corrected to account for the multiple comparisons, not one of them was statistically significant.

Further studies have suggested that the S-phase fraction is indeed related to tumor prognosis, but the evidence was poor for many years. The method continued to be used in cancer studies for several years after its flaws were publicized, and a 2005 set of reporting guidelines for cancer prognostic factor studies noted the following: "Despite years of research and hundreds of reports on tumor markers in oncology, the number of markers that have emerged as clinically useful is pitifully small."[2] Apart from poor statistical power, incomplete reporting of results, and sampling biases, the choice of "optimal" cut points was cited as a key reason for this problem.

Statistical Brownout

A major objection to dichotomization is that it throws away information. Instead of using a precise number for every patient or observation, you split observations into groups and throw away the numbers. This reduces the statistical power of your study—a major problem when so many studies are already underpowered. You'll get less precise estimates of the correlations you're trying to measure and will often underestimate effect sizes. In general, this loss of power and precision is the same you'd get by throwing away a third of your data.[3]

Let's go back to the example study measuring the health impacts of obesity. Say you split patients into "normal" and "overweight" groups based on their *body mass index*, taking a BMI of 25 to be the maximum for the normal range. (This is the standard cutoff used in clinical practice.) But then you've lost the distinction between all BMIs above this cutoff. If the heart-disease rate increases with weight, it's much more difficult to tell *how much* it increases because you didn't record the difference between, say, mildly overweight and morbidly obese patients.

To put this another way, imagine if the "normal" group consisted of patients with BMIs of exactly 24, while the "overweight" group had BMIs of 26. A major difference between the groups would be surprising since they're not very different. On the other hand, if the "overweight" group all had BMIs of 36, a major difference would be much less surprising and indicate a much smaller difference per BMI unit. Dichotomization eliminates this distinction, dropping useful information and statistical power.

Perhaps it was a silly choice to use only two groups—what about underweight patients?—but increasing the number of groups means the number of patients in each group decreases. More groups might produce a more detailed analysis, but the heart disease rate estimates for each group will be based on less data and have wider confidence intervals. And splitting data into more groups means making more decisions about *where* to split the data, making different studies yet more difficult to compare and making it even easier for researchers to generate false positives.

Confounded Confounding

You may wonder the following: if I have enough data to achieve statistical significance after I've dichotomized my data, does the dichotomization matter? As long as I can make up for the lost statistical power with extra data, why not dichotomize to make the statistical analysis easy?

That's a legitimate argument. But analyzing data without dichotomizing isn't that hard. Regression analysis is a common procedure, supported by nearly every statistical software package and covered in numerous books. Regression doesn't involve dichotomization—it uses the full data, so there is no cutoff to choose and no loss of statistical power. So why water down your data? But more importantly, dichotomization does more than cut power. Counterintuitively, it also introduces false positives.

We are often interested in controlling for confounding factors. You might measure two or three variables (or two or three dozen) along with the outcome variable and attempt to determine the unique effect of each variable on the outcome after the other variables have been "controlled for." If you have two variables and one outcome, you could easily do this by dichotomizing the two variables and using a two-way analysis of variance (ANOVA) table, a simple, commonly performed procedure supported by every major statistical software package.

Unfortunately, the worst that could happen isn't a false negative. By dichotomizing and throwing away information, you eliminate the ability to distinguish between confounding factors.[4]

Consider an example. Say you're measuring the effect of a number of variables on the quality of health care a person receives. Health-care quality (perhaps measured using a survey) is the outcome variable. For predictor variables, you use two measurements: the subject's personal net worth in dollars and the length of the subject's personal yacht.

You would expect a good statistical procedure to deduce that wealth impacts quality of health care but yacht size does not. Even though yacht size and wealth tend to increase together, it's not your yacht that gets you better health care. With enough data, you would notice that people of the same wealth can have differently sized yachts—or no yachts at all—but still get a similar quality of care. This indicates that wealth is the primary factor, not yacht length.

But by dichotomizing the variables, you've effectively cut the data down to four points. Each predictor can be only "above the median" or "below the median," and no further information is recorded. You no longer have the data needed to realize that yacht length has nothing to do with health care. As a result, the ANOVA procedure falsely claims that yachts and health care are related. Worse, this false correlation isn't statistically significant only 5% of the time—from the ANOVA's perspective, it's a *true* correlation, and it is detected as often as the statistical power of the test allows it.

Of course, you could have figured out that yacht size wouldn't matter, even without data. You could have left it out of the analysis and saved a lot of trouble. But you don't usually know in advance which variables are most important—you depend on your statistical analysis to tell you.

Regression procedures can easily fit this data without any dichotomization, while producing false-positive correlations only at the rate you'd expect. (Of course, as the correlation between wealth and yacht size becomes stronger, it becomes more difficult to distinguish between their effects.) While the mathematical theory of regression with multiple variables can be more advanced than many practicing scientists care to understand, involving a great deal of linear algebra, the basic concepts and results are easy to understand and interpret. There's no good reason not to use it.

TIPS • Don't arbitrarily split continuous variables into discrete groups unless you have good reason. Use a statistical procedure that can take full advantage of the continuous variables.

• If you do need to split continuous variables into groups for some reason, don't choose the groups to maximize your statistical significance. Define the split in advance, use the same split as in previous similar research, or use outside standards (such as a medical definition of obesity or high blood pressure) instead.

8

MODEL ABUSE

Let's move on to regression. Regression in its simplest form is fitting a straight line to data: finding the equation of the line that best predicts the outcome from the data. With this equation, you can use a measurement, such as body mass index, to predict an outcome like blood pressure or medical costs.

Usually regression uses more than one predictor variable. Instead of just body mass index, you might add age, gender, amount of regular exercise, and so on. Once you collect medical data from a representative sample of patients, the regression procedure would use the data to find the best equation to represent the relationship between the predictors and the outcome.

As we saw in Chapter 7, regression with multiple variables allows you to *control for* confounding factors in a study. For example, you might study the impact of class size on students' performance on standardized tests, hypothesizing that smaller classes improve test scores. You could use regression to find the relationship between size and score, thus testing whether test scores rise as class size falls—but there's a *confounding variable.*

If you find a relationship, then perhaps you've shown that class size is the cause, but the cause could also be another factor that influences class size and scores together. Perhaps schools with bigger budgets can afford more teachers, and hence smaller classes, and can also afford more books, higher teacher salaries, more support staff, better science labs, and other resources that help students learn. Class size could have nothing to do with it.

To control for the confounding variable, you record each school's total budget and include it in your regression equation, thus separating the effect of budget from the effect of class size. If you examine schools with similar budgets and different class sizes, regression produces an equation that lets us say, "For schools *with the same budget,* increasing class size by one student lowers test scores by this many points." The confounding variable is hence controlled for. Of course, there may be confounding variables you aren't aware of or don't know how to measure, and these could influence your results; only a truly randomized experiment eliminates all confounding variables.

There are many more versions of regression than the simple one presented here. Often the relationship between two variables isn't a simple linear equation. Or perhaps the outcome variable isn't quantitative, like blood pressure or a test score, but categorical. Maybe you want to predict whether a patient will suffer complications after a surgery, using his or her age, blood pressure, and other vital signs. There are many varieties of procedures to account for these possibilities.

All kinds of regression procedures are subject to common problems. Let's start with the simplest problem: overfitting, which is the result of excessive enthusiasm in data analysis.

Fitting Data to Watermelons

A common watermelon selection strategy is to knock on the melons and pick those with a particularly hollow sound, which apparently results from desirable characteristics of watermelon flesh. With the right measurement equipment, it should be possible to use statistics to find an algorithm that can predict the ripeness of any melon from its sound.

I am particularly interested in this problem because I once tried to investigate it, building a circuit to connect a fancy accelerometer to my computer so I could record the thump of watermelons. But I tested only eight melons—not nearly enough data to build an accurate ripeness-prediction system. So I was understandably excited when I came across a paper that claimed to predict watermelon ripeness with fantastic accuracy: acoustic measurements could predict 99.9% of the variation in ripeness.[1]

But let's think. In this study, panelists tasted and rated 43 watermelons using a five-point ripeness scale. Regression was used to predict the ripeness rating from various acoustic measurements. How could the regression equation's accuracy be so high? If you had the panelists rerate the melons, they probably wouldn't agree with *their own ratings* with 99.9% accuracy. Subjective ratings aren't that consistent. No procedure, no matter how sophisticated, could predict them with such accuracy.

Something is wrong. Let's evaluate their methods more carefully.

Each watermelon was vibrated at a range of frequencies, from 1 to 1,000 hertz, and the phase shift (essentially, how long it took the vibration to travel through the melon) was measured at each frequency. There were 1,600 tested frequencies, so there were 1,600 variables in the regression model. Each one's relationship to ripeness has to be estimated.

Now, with more variables than watermelons, I could fit a *perfect* regression model. Just like a straight line can be made to fit perfectly between any two data points, an equation with 43 variables can be used to perfectly fit the measurements of 43 melons. This is serious overkill. Even if there is no relationship whatsoever between acoustics and ripeness, I can fit a regression equation that gives 100% accuracy on the 43 watermelons. It will account for not just the true relationship between acoustics and ripeness (if one exists) but also random variation in individual ratings and measurements. I will believe the model fits perfectly—but tested on new watermelons with their own measurement errors and subjective ratings, it may be useless.

The authors of the study attempted to sidestep this problem by using stepwise regression, a common procedure for selecting which variables are the most important in a regression. In its simplest form, it goes like this: start by using none of the

1,600 frequency measurements. Perform 1,600 hypothesis tests to determine which of the frequencies has the most statistically significant relationship with the outcome. Add that frequency and then repeat with the remaining 1,599. Continue the procedure until there are no statistically significant frequencies.

Stepwise regression is common in many scientific fields, but it's usually a bad idea.[2] You probably already noticed one problem: multiple comparisons. Hypothetically, by adding only statistically significant variables, you avoid overfitting, but running so many significance tests is bound to produce false positives, so some of the variables you select will be bogus. Stepwise regression procedures provide no guarantees about the overall false positive rate, nor are they guaranteed to select the "best" combination of variables, however you define "best." (Alternative stepwise procedures use other criteria instead of statistical significance but suffer from many of the same problems.)

So despite the veneer of statistical significance, stepwise regression is susceptible to egregious *overfitting*, producing an equation that fits the data nearly perfectly but that may prove useless when tested on a separate dataset. As a test, I simulated random watermelon measurements with absolutely zero correlation with ripeness, and nonetheless stepwise regression fit the data with 99.9% accuracy. With so many variables to choose from, it would be more surprising if it didn't.

Most uses of stepwise regression are not in such extreme cases. Having 1,600 variables to choose from is extraordinarily rare. But even in modest cases with 100 observations of a few dozen variables, stepwise regression produces inflated estimates of accuracy and statistical significance.[3,4]

Truth inflation is a more insidious problem. Remember, "statistically insignificant" does not mean "has no effect whatsoever." If your study is underpowered—you have too many variables to choose from and too little data—then you may not have enough data to reliably distinguish each variable's effect from zero. You'll include variables only if you are unlucky enough to overestimate their effect on the outcome. Your model will be heavily biased. (Even when not using a formal stepwise regression procedure, it's common practice to throw out "insignificant" variables to simplify a model, leading to the same problem.)

There are several variations of stepwise regression. The version I just described is called *forward selection* since it starts from scratch and starts including variables. The alternative, *backward elimination*, starts by including all 1,600 variables and excludes those that are statistically insignificant, one at a time. (This would fail, in this case: with 1,600 variables but only 43 melons, there isn't enough data to uniquely determine the effects of all 1,600 variables. You would get stuck on the first step.) It's also possible to change the criteria used to include new variables; instead of statistical significance, more-modern procedures use metrics like the Akaike information criterion and the Bayesian information criterion, which reduce overfitting by penalizing models with more variables. Other variations add and remove variables at each step according to various criteria. None of these variations is guaranteed to arrive at the same answer, so two analyses of the same data could arrive at very different results.

For the watermelon study, these factors combined to produce implausibly accurate results. How can a regression model be fairly evaluated, avoiding these problems? One option is *cross-validation*: fit the model using only a portion of the melons and then test its effectiveness at predicting the ripeness of the other melons. If the model overfits, it will perform poorly during cross-validation. One common cross-validation method is *leave-out-one cross-validation*, where the model is fit using all but one data point and then evaluated on its ability to predict that point; the procedure is repeated with each data point left out in turn. The watermelon study claims to have performed leave-out-one cross-validation but obtained similarly implausible results. Without access to the data, I'm not sure whether the method genuinely works.

Despite these drawbacks, stepwise regression continues to be popular. It's an intuitively appealing algorithm: select the variables with statistically significant effects. But choosing a single model is usually foolishly overconfident. With so many variables to choose from, there are often many combinations of variables that predict the outcome nearly as well. Had I picked 43 more watermelons to test, I probably would have selected a different subset of the 1,600 possible acoustic predictors of ripeness. Stepwise regression produces misleading certainty—the claim that these 20 or 30 variables are "the" predictors of ripeness, though dozens of others could do the job.

Of course, in some cases there may be a good reason to believe that only a few of the variables have any effect on the outcome. Perhaps you're identifying the genes responsible for a rare cancer, and though you have thousands of candidates, you know only a few are the cause. Now you're not interested in making the best predictions—you just want to identify the responsible genes. Stepwise regression is still not the best tool;

the lasso (short for *least absolute shrinkage and selection operator,* an inspired acronym) has better mathematical properties and doesn't fool the user with claims of statistical significance. But the lasso is not bulletproof, and there is no perfect automated solution.

Correlation and Causation

When you have used multiple regression to model some outcome—like the probability that a given person will suffer a heart attack, given that person's weight, cholesterol, and so on—it's tempting to interpret each variable on its own. You might survey thousands of people, asking whether they've had a heart attack and then doing a thorough physical examination, and produce a model. Then you use this model to give health advice: lose some weight, you say, and make sure your cholesterol levels fall within this healthy range. Follow these instructions, and your heart attack risk will decrease by 30%!

But that's not what your model says. The model says that people with cholesterol and weight within that range have a 30% lower risk of heart attack; it *doesn't* say that if you put an overweight person on a diet and exercise routine, that person will be less likely to have a heart attack. You didn't collect data on that! You didn't intervene and change the weight and cholesterol levels of your volunteers to see what would happen.

There could be a confounding variable here. Perhaps obesity and high cholesterol levels are merely symptoms of some other factor that also causes heart attacks; exercise and statin pills may fix them but perhaps not the heart attacks. The regression model says lower cholesterol means fewer heart attacks, but that's correlation, not causation.

One example of this problem occurred in a 2010 trial testing whether omega-3 fatty acids, found in fish oil and commonly sold as a health supplement, can reduce the risk of heart attacks. The claim that omega-3 fatty acids reduce heart attack risk was supported by several observational studies, along with some experimental data. Fatty acids have anti-inflammatory properties and can reduce the level of triglycerides in the bloodstream—two qualities known to correlate with reduced heart attack risk. So it was reasoned that omega-3 fatty acids should reduce heart attack risk.[5]

But the evidence was observational. Patients with low triglyceride levels had fewer heart problems, and fish oils reduce triglyceride levels, so it was spuriously concluded that fish oil should protect against heart problems. Only in 2013 was a large randomized controlled trial published, in which patients were given either fish oil or a placebo (olive oil) and monitored for five years. There was no evidence of a beneficial effect of fish oil.[6]

Another problem arises when you control for multiple confounding factors. It's common to interpret the results by saying, "If weight increases by one pound, with all other variables held constant, then heart attack rates increase by . . ." Perhaps that is true, but it may not be *possible* to hold all other variables constant in practice. You can always quote the numbers from the regression equation, but in reality the act of gaining a pound of weight also involves other changes. Nobody ever gains a pound with all other variables held constant, so your regression equation doesn't translate to reality.

Simpson's Paradox

When statisticians are asked for an interesting paradoxical result in statistics, they often turn to Simpson's paradox.[*] *Simpson's paradox* arises whenever an apparent trend in data, caused by a confounding variable, can be eliminated or reversed by splitting the data into natural groups. There are many examples of the paradox, so let me start with the most popular.

In 1973, the University of California, Berkeley, received 12,763 applications for graduate study. In that year's admissions process, 44% of male applicants were accepted but only 35% of female applicants were. The university administration, fearing a gender discrimination lawsuit, asked several of its faculty to take a closer look at the data.[†]

Graduate admissions, unlike undergraduate admissions, are handled by each academic department independently. The initial investigation led to a paradoxical conclusion: of 101 separate graduate departments at Berkeley, only 4 departments showed a statistically significant bias against admitting women. At the same time, six departments showed a bias against *men*,

---

[*]Simpson's paradox was discovered by Karl Pearson and Udny Yule and is thus an example of Stigler's law of eponymy, discovered by Robert Merton, which states that no scientific discovery is named after the original discoverer.
[†]The standard version of this story claims that the university was sued for discrimination, but nobody ever says who filed the suit or what became of it. A *Wall Street Journal* interview with a statistician involved in the original investigation reveals that the lawsuit never happened.[7] The mere fear of a lawsuit was sufficient to trigger an investigation. But the lawsuit story has been around so long that it's commonly regarded as fact.

which was more than enough to cancel out the deficit of women caused by the other four departments.

How could Berkeley as a whole appear biased against women when individual departments were generally not? It turns out that men and women did not apply to all departments in equal proportion. For example, nearly two-thirds of the applicants to the English department were women, while only 2% of mechanical engineering applicants were. Furthermore, some graduate departments were more selective than others.

These two factors accounted for the perceived bias. Women tended to apply to departments with many qualified applicants and little funding, while men applied to departments with fewer applicants and surpluses of research grants. The bias was not at Berkeley, where individual departments were generally fair, but further back in the educational process, where women were being shunted into fields of study with fewer graduate opportunities.[8]

Simpson's paradox came up again in a 1986 study on surgical techniques to remove kidney stones. An analysis of hundreds of medical records appeared to show that percutaneous nephrolithotomy, a minimally invasive new procedure for removing kidney stones, had a higher success rate than traditional open surgery: 83% instead of 78%.

On closer inspection, the trend reversed. When the data was split into small and large kidney-stone groups, percutaneous nephrolithotomy performed *worse* in both groups, as shown in Table 8-1. How was this possible?

Table 8-1: Success Rates for Kidney Stone Removal Surgeries

Treatment	Diameter < 2 cm	Dia. ≥ 2 cm	Overall
Open surgery	93%	73%	78%
Percutaneous nephrolithotomy	87%	69%	83%

The problem was that the study did not use randomized assignment. It was merely a review of medical records, and it turned out that doctors were systematically biased in how they treated each patient. Patients with large, difficult-to-remove kidney stones underwent open surgery, while those with small, easy-to-remove stones had the nephrolithotomy.[9] Presumably, doctors were more comfortable using the new, unfamiliar procedure on patients with small stones and reverted to open surgery for tough cases.

The new surgery wasn't necessarily better but was tested on the easiest patients. Had the surgical method been chosen by random assignment instead of at the surgeon's discretion, there'd have been no such bias. In general, random assignment eliminates confounding variables and prevents Simpson's paradox from giving us backward results. Purely observational studies are particularly susceptible to the paradox.

This problem is common in medicine, as illustrated by another example. Bacterial meningitis is an infection of tissues surrounding the brain and spinal cord and is known to progress quickly and cause permanent damage if not immediately treated, particularly in children. In the United Kingdom, general practitioners typically administer penicillin to children they believe have meningitis before sending them to the hospital for further tests and treatment. The goal is to start treatment as soon as possible, without waiting for the child to travel to the hospital.

To see whether this early treatment was truly beneficial, an observational study examined records of 448 children diagnosed with meningitis and admitted to the hospital. Simple analysis showed that children given penicillin by general practitioners were less likely to die in treatment.

A more careful look at the data reversed this trend. Many children had been admitted directly to the hospital and never saw a general practitioner, meaning they didn't receive the initial penicillin shot. They were also the children with the most severe illnesses—the children whose parents rushed them directly to the hospital. What if they are excluded from the data and you ask only, "Among children who saw their general practitioner first, did those administered penicillin have better outcomes?" Then the answer is an emphatic *no*. The children administered penicillin were much more likely to die.[10]

But this was an observational study, so you can't be sure the penicillin *caused* their deaths. It's hypothesized that toxins released during the destruction of the bacteria could cause shock, but this has not been experimentally proven. Or perhaps general practitioners gave penicillin only to children who had the most severe cases. You can't be sure without a randomized trial.

Unfortunately, randomized controlled experiments are difficult and sometimes impossible to run. For example, it may be considered unethical to deliberately withhold penicillin from children with meningitis. For a nonmedical example, if you compare flight delays between United Airlines and Continental Airlines, you'll find United has more flights delayed on average. But at each individual airport in the comparison, Continental's flights are more likely to be delayed. It turns out United operates more flights out of cities with poor weather. Its average is dragged down by the airports with the most delays.[7]

But you can't randomly assign airline flights to United or Continental. You can't always eliminate every confounding factor. You can only measure them and hope you've measured them all.

TIPS • Remember that a statistically insignificant variable does not necessarily have zero effect; you may not have the power needed to detect its effect.

• Avoid stepwise regression when possible. Sometimes it's useful, but the final model is biased and difficult to interpret. Other selection techniques, such as the lasso, may be more appropriate. Or there may be no need to do variable selection at all.

• To test how well your model fits the data, use a separate dataset or a procedure such as cross-validation.

• Watch out for confounding variables that could cause misleading or reversed results, as in Simpson's paradox, and use random assignment to eliminate them whenever possible.

9

RESEARCHER FREEDOM: GOOD VIBRATIONS?

There's a common misconception that statistics is boring and monotonous. Collect lots of data; plug numbers into Excel, SPSS, or R; and beat the software with a stick until it produces colorful charts and graphs. Done! All the statistician must do is enter some commands and read the results.

But one must choose *which* commands to use. Two researchers attempting to answer the same question can and often do perform entirely different statistical analyses. There are many decisions to make.

What do I measure?

This isn't as obvious as it sounds. If I'm testing a psychiatric medication, I could use several different scales to measure symptoms: various brain function tests, reports from doctors, or all sorts of other measurements. Which will be most useful?

Which variables do I adjust for?

In a medical trial, I might control for patient age, gender, weight, BMI, medical history, smoking, or drug use, or for the results of medical tests done before the start of the study. Which of these factors are important? Which can be ignored? How do I measure them?

Which cases do I exclude?

If I'm testing diet plans, maybe I want to exclude test subjects who came down with diarrhea during the trial, since their results will be abnormal. Or maybe diarrhea is a side effect of the diet and I must include it. There will always be some results that are out of the ordinary, for reasons known or unknown. I may want to exclude them or analyze them specially. Which cases count as outliers? What do I do with them?

How do I define groups?

For example, I may want to split patients into "overweight," "normal," and "underweight" groups. Where do I draw the lines? What do I do with a muscular bodybuilder whose BMI is in the "overweight" range?

What about missing data?

Perhaps I'm testing cancer remission rates with a new drug. I run the trial for five years, but some patients will have tumors reappear after six years or eight years. My data does not include their recurrence. Or perhaps some patients dropped out because of side effects or personal problems. How do I account for this when measuring the effectiveness of the drug?

How much data should I collect?

Should I stop when I have a definitive result or continue as planned until I've collected all the data? What if I have trouble enrolling as many patients as desired?

It can take hours of exploration to see which procedures are most appropriate. Papers usually explain the statistical analysis performed but don't always explain why researchers chose one method over another or what the results would have been had they chosen a different method. Researchers are free to choose whatever methods they feel appropriate—and though they may make good choices, what happens if they analyze the data differently?

This statistical freedom allows bias to creep into analysis undetected, even when analysts have the best of intentions. A few analysis decisions can change results dramatically, suggesting that perhaps analysts should make the decisions *before* they see the data. Let's start with the outsized impact of small analysis decisions.

A Little Freedom Is a Dangerous Thing

In simulations, it's possible to get effect sizes different by a factor of two simply by adjusting for different variables, excluding different sets of cases, and handling outliers differently.[1] Even reasonable practices, such as remeasuring patients with strange laboratory test results or removing clearly abnormal patients, can bring a statistically insignificant result to significance.[2] Apparently, being free to analyze how you want gives you enormous control over your results!

A group of researchers demonstrated this phenomenon with a simple experiment. Twenty undergraduates were randomly assigned to listen to either "When I'm Sixty-Four" by the Beatles or "Kalimba," a song that comes with the Windows 7 operating system. Afterward, they were asked their age and their father's age. The two groups were compared, and it was found that "When I'm Sixty-Four" listeners were a year and a half younger on average, controlling for their father's age, with $p < 0.05$. Since the groups were randomly assigned, the only plausible source of the difference was the music.

Rather than publishing *The Musical Guide to Staying Young*, the researchers explained the tricks they used to obtain this result. They didn't decide in advance how much data to collect; instead, they recruited students and ran statistical tests periodically to see whether a significant result had been achieved. (You saw earlier that such stopping rules can inflate false-positive rates significantly.) They also didn't decide in advance to control for the age of the subjects' fathers, instead asking how old they *felt*, how much they would enjoy eating at a diner, the square root of 100, their mother's age, their agreement with "computers are complicated machines," whether they would take advantage of an early-bird special, their political orientation, which of four Canadian quarterbacks they believed won an award, how often they refer to the past as "the good old days," and their gender.

Only after looking at the data did the researchers decide on which outcome variable to use and which variables to control for. (Had the results been different, they might have reported that "When I'm Sixty-Four" causes students to, say, be less able to calculate the square root of 100, controlling for their knowledge of Canadian football.) Naturally, this freedom allowed the researchers to make multiple comparisons and inflated their false-positive rate. In a published paper, they wouldn't need to mention the other insignificant variables; they'd be free to discuss the apparent antiaging benefit of the Beatles. The fallacy would not be visible to the reader.

Further simulation by the researchers suggested that if scientists try different statistical analyses until one works—say, by controlling for different combinations of variables and trying different sample sizes—false positive rates can jump to more than 50% for a given dataset.[3]

This example sounds outlandish, and most scientists would protest that they don't intentionally tinker with the data until a significant result appears. They construct a hypothesis, collect data, explore the data a bit, and run a reasonable statistical analysis to test the hypothesis. Perhaps we could have tried 100 analyses until we got a fantastic result, they say, but we didn't. We picked one analysis that seemed appropriate for the data and stuck with it.

But the choice of analysis strategy is always based on the data. We look at our data to decide which variables to include, which outliers to remove, which statistical tests to use, and which outcomes to examine. We do this not with the explicit goal of finding the most statistically significant result but to design an analysis that accounts for the peculiarities that arise in any dataset. Had we collected different data—had that one patient suffered from chronic constipation instead of acute diarrhea—we would choose a different statistical analysis. We bias the analysis to produce results that "make sense."

Furthermore, a single prespecified scientific hypothesis does not necessarily correspond to a single *statistical* hypothesis. Many different statistical results could all be interpreted to support a hypothesis. You may believe that one drug has fewer side effects than another, but you will accept statistically significant drops in any of a dozen side effects as evidence. You may believe that women are more likely to wear red or pink during ovulation, but you will accept statistically significant effects for red shirts, pink shirts, or the combination of both.* (Or perhaps you will accept effects for shirts, pants, hats, socks, or other kinds of clothing.) If you hypothesize that ovulation makes single women more liberal, you will accept changes in any of their voting choices, religious beliefs, and political values as evidence.† The choices that produce interesting results will attract our attention and engage our human tendency to build plausible stories for any outcome.

The most worrying consequence of this statistical freedom is that researchers may unintentionally choose the statistical analysis most favorable to them. Their resulting estimates of uncertainty—standard errors, confidence intervals, and so on—will be biased. The false-positive rate will be inflated because the data guided their statistical design.

Avoiding Bias

In physics, unconscious biases have long been recognized as a problem. Measurements of physical constants, such as the speed of light or subatomic particle properties, tend to cluster around previous measurements rather than the eventually accepted "truth."[8] It seems an experimentalist, obtaining results that disagree with earlier studies, "searches for the source or sources of such errors, and continues to search until he gets a result close to the accepted value. *Then he stops!*"[9]

Seeking to eliminate this bias, particle physicists have begun performing *blind analyses*: the scientists analyzing the data avoid calculating the value of interest until after the analysis procedure is finalized. Sometimes this is easy: Frank Dunnington, measuring the electron's charge-to-mass ratio in the early 1930s, had his machinist build the experimental apparatus with the detector close to, but not exactly at, the optimal angle. Without the precise angle measurement, Dunnington could not calculate his final answer, so he devised his analysis procedures while unable to subconsciously bias the results. Once he was ready, he measured the angle and calculated the final ratio.

---

*This was a real study, claiming women at peak fertility were three times more likely to wear red or pink.[4] Columbia University statistician Andrew Gelman wrote an article in *Slate* criticizing the many degrees of freedom in the study, using it as an example to attack statistical methods in psychology in general.[5]
†I am not making up this study either. It also found that "ovulation led married women to become more conservative."[6] A large replication attempt found no evidence for either claim.[7]

Blind analysis isn't always this straightforward, of course, but particle physicists have begun to adopt it for major experiments. Other blinding techniques include adding a constant to all measurements, keeping this constant hidden from analysts until the analysis is finalized; having independent groups perform separate parts of the analysis and only later combining their results; or using simulations to inject false data that is later removed. Results are unblinded only after the research group is satisfied that the analysis is complete and appropriate.

In some medical studies, triple blinding is performed as a form of blind analysis; the patients, doctors, and statisticians all do not know which group is the control group until the analysis is complete. This does not eliminate all sources of bias. For example, the statistician may not be able to unconsciously favor the treatment group, but she may be biased toward a larger difference between groups. More extensive blinding techniques are not in frequent use, and significant methodological research is required to determine how common statistical techniques can be blinded without making analysis impractical.

Instead of triple blinding, one option is to remove the statistician's freedom of choice. A limited form of this, covering the design and execution of the experiment rather than its analysis, is common in medicine. Doctors are required to draft a clinical trial protocol explaining how the data will be collected, including the planned sample size and measured outcome variables, and then the protocol is reviewed by an ethics committee to ensure it adequately protects patient safety and privacy. Because the protocol is drafted before data is collected, doctors can't easily tinker with the design to obtain favorable results. Unfortunately, many studies depart from their protocols, allowing for researcher bias to creep in.[10,11] Journal editors often don't compare submitted papers to original protocols and don't require authors to explain why their protocols were violated, so there is no way to determine the motivation for the changes.

Many scientific fields have no protocol publication requirement, and in sciences such as psychology, psychiatry, and sociology, there is often no single agreed-upon methodology to use for a particular experiment. Appropriate designs for medical trials or physics experiments have been analyzed to death, but it's often unclear how to handle less straightforward behavioral studies. The result is an explosion of diversity in study design, with every new paper using a different combination of methods. When there is intense pressure to produce novel results, as there usually is in the United States, researchers in these fields tend to produce biased and extreme results more frequently because of their freedom in experimental design and data analysis.[12] In response, some have proposed allowing protocol registration for confirmatory research, lending subsequent results greater credibility.

Of course, to paraphrase Helmuth von Moltke, no analysis plan survives contact with the data. There may be complications and problems you did not anticipate. Your assumptions about the distribution of measurements, the correlation between variables, and the likely causes of outliers—all essential to your choice of analysis—may be entirely wrong. You might have no idea what assumptions to make before collecting the data. When that happens, it's better to correct your analysis than to proceed with an obviously wrong preplanned analysis.

It may not even be possible to prespecify an analysis before seeing the data. Perhaps you decide to test a new hypothesis using a common dataset that you have used for years, perhaps you aren't sure what hypothesis is relevant until you see the data, or perhaps the data suggests interesting hypotheses you hadn't thought of before collecting it. For some fields, prepublication replication can solve this problem: collect a new, independent dataset and analyze it using exactly the same methods. If the effect remains, you can be confident in your results. (Be sure your new sample has adequate statistical power.) But for economists studying a market crash, it's not possible (or at least not ethical) to arrange for another one. For a doctor studying a cancer treatment, patients may not be able to wait for replication.

The proliferation of statistical techniques has given us useful tools, but it seems they've been put to use as blunt objects with which to beat the data until it confesses. With preregistered analyses, blinding, and further research into experimental methods, we can start to treat our data more humanely.

TIPS
- Before collecting data, plan your data analysis, accounting for multiple comparisons and including any effects you'd like to look for.

- Register your clinical trial protocol if applicable.

- If you deviate from your planned protocol, note this in your paper and provide an explanation.

- Don't just torture the data until it confesses. Have a specific statistical hypothesis in mind before you begin your analysis.

10

EVERYBODY MAKES MISTAKES

Until now, I have presumed that scientists are capable of making statistical computations with perfect accuracy and err only in their choice of appropriate numbers to compute. Scientists may misuse the results of statistical tests or fail to make relevant computations, but they can at least calculate a *p* value, right?

Perhaps not.

Surveys of statistically significant results reported in medical and psychological trials suggest that many *p* values are wrong and some statistically insignificant results are actually significant when computed correctly.[1,2] Even the prestigious journal *Nature* isn't perfect, with roughly 38% of papers making typos and calculation errors in their *p* values.[3] Other reviews find examples of misclassified data, erroneous duplication of data, inclusion of the wrong dataset entirely, and other mix-ups, all concealed by papers that did not describe their analysis in enough detail for the errors to be easily noticed.[4]

These sorts of mistakes are to be expected. Scientists may be superhumanly caffeinated, but they're still human, and the constant pressure to publish means that thorough documentation and replication are ignored. There's no incentive for researchers to make their data and calculations available for inspection or to devote time to replicating other researchers' results.

As these problems have become more widely known, software tools have advanced to make analysis steps easier to record and share. Scientists have yet to widely adopt these tools, however, and without them, thoroughly checking work can be a painstaking process, as illustrated by a famous debacle in genetics.

Irreproducible Genetics

The problems began in 2006, when a new genetic test promised to allow chemotherapy treatments to be carefully targeted to the patient's specific variant of cancer. Duke University researchers ran trials indicating that their technique could determine which drugs a tumor would be most sensitive to, sparing patients the side effects of ineffective treatments. Oncologists were excited at the prospect, and other researchers began their own studies. But first they asked two biostatisticians, Keith Baggerly and Kevin Coombes, to check the data.

This was more difficult than they expected. The original papers did not give sufficient detail to replicate the analysis, so Baggerly and Coombes corresponded with the Duke researchers to get raw data and more details. Soon they discovered problems. Some of the data was mislabeled—groups of cells that were resistant to a drug were marked as sensitive instead, and vice versa. Some samples were duplicated in the data, sometimes marked as both sensitive and resistant. A correction issued by the Duke researchers fixed some of these issues but introduced more duplicated data at the same time. Some data was accidentally shifted by one so that measurements from one set of cells were used when analyzing a different cell line. Genetic microarrays, which I discussed earlier in the context of pseudoreplication, varied significantly between batches, and the effect of the microarray equipment could not be separated from the true biological differences. Figures allegedly showing results for one drug actually contained the results for a different drug.

In short, the research was a mess.[5] Despite many of the errors being brought to the attention of the Duke researchers, several clinical trials using the genetic results began, funded by the National Cancer Institute. Baggerly and Coombes attempted to publish their responses to the research in the same academic journals that published the original research, but in several cases they were rejected—groundbreaking research is more interesting than tedious statistical detail. Nonetheless, the National Cancer Institute caught wind of the problems and asked Duke administrators to review the work. The university responded by creating an external review committee that had no access to Baggerly and Coombes' results. Unsurprisingly, they found no errors, and the trials continued.[6]

The errors attracted serious attention only later, some time after Baggerly and Coombes published their discoveries, when a trade magazine reported that the lead Duke researcher, Anil Potti, had falsified his résumé. Several of his papers were retracted, and Potti eventually resigned from Duke amid accusations of fraud. Several trials using the results were stopped, and a company set up to sell the technology closed.[7]

The Potti case illustrates two problems: the lack of reproducibility in much of modern science and the difficulty of publishing negative and contradictory results in academic journals. I'll save the latter issue for the next chapter. Reproducibility has become a popular buzzword, and you can probably see why: Baggerly and Coombes estimate they spent 2,000 hours figuring out what Potti had done and what went wrong. Few academics have that kind of spare time. If Potti's analysis software and data were openly available for inspection, skeptical colleagues would not be forced to painstakingly reconstruct every step of his work—they could simply read through the code and see where every chart and graph came from.

The problem was not just that Potti did not share his data readily. Scientists often do not record and document the steps they take converting raw data to results, except in the often-vague form of a scientific paper or whatever is written down in a lab notebook. Raw data has to be edited, converted to other formats, and linked with other datasets; statistical analysis has to be performed, sometimes with custom software; and plots and tables have to be created from the results. This is often done by hand, with bits of data copied and pasted into different data files and spreadsheets—a tremendously error-prone process. There is usually no definitive record of these steps apart from the overstressed memory of the graduate student responsible, though we would like to be able to examine and reproduce every step of the process years after the student has graduated.

Making Reproducibility Easy

Ideally, these steps would be *reproducible*: fully automated, with the computer source code available for inspection as a definitive record of the work. Errors would be easy to spot and correct, and any scientist could download the dataset and code and produce exactly the same results. Even better, the code would be combined with a description of its purpose.

Statistical software has been advancing to make this possible. A tool called Sweave, for instance, makes it easy to embed statistical analyses performed using the popular R programming language inside papers written in LaTeX, a typesetting system commonly used for scientific and mathematical publications. The result looks just like any scientific paper, but another scientist reading the paper and curious about its methods can download the source code, which shows exactly how all the numbers and plots were calculated. But academic journals, which use complicated typesetting and publishing systems, do not yet accept Sweave publications, so its use is limited.

Similar tools are emerging for other programming languages. Data analysts using the Python programming language, for example, can record their progress using the IPython Notebook, which weaves together text descriptions, Python code, and plots and graphics generated by the Python code. An IPython Notebook can read like a narrative of the analysis process, explaining how data is read in, processed, filtered, analyzed, and plotted, with code accompanying the text. An error in any step can be corrected and the code rerun to obtain new results. And notebooks can be turned into web pages or LaTeX documents, so other researchers don't need to install IPython to read the code. Best of all, the IPython Notebook system has been extended to work with other languages, such as R.

Journals in heavily computational fields, such as computational biology and statistics, have begun adopting code-sharing policies encouraging public posting of analysis source code. These policies have not yet been as widely adopted as data-sharing policies, but they are becoming more common.[8] A more comprehensive strategy to ensure reproducibility and ease of error detection would follow the "Ten Simple Rules for Reproducible Computational Research," developed by a group of biomedical researchers.[9] These rules include automating data manipulation and reformatting, recording all changes to analysis software and custom programs using a software version control system, storing all raw data, and making all scripts and data available for public analysis. Every scientist has experienced the confusion of reading a paper and wondering, "How the hell did they get *that* number?", and these rules would make that question much easier to answer.

That's quite a lot of work, with little motivation for the scientist, who already knows how the analysis was done. Why spend so much time making code suitable for *other* people to benefit from, instead of doing more research? There are many advantages. Automated data analysis makes it easy to try software on new datasets or test that each piece functions correctly. Using a version control system means you have a record of every change, so you're never stuck wondering, "How could this code have worked last Tuesday but not now?" And a comprehensive record of calculations and code means you can always redo it later; I was once very embarrassed when I had to reformat figures in a paper for publication, only to realize that I didn't remember what data I had used to make them. My messy analysis cost me a day of panic as I tried to re-create the plots.

But even if they *have* fully automated their analysis, scientists are understandably reluctant to share their code. What if a competing scientist uses it to beat you to a discovery? Since they aren't required to disclose their code, they don't have to disclose that they used yours; they can get academic credit for a discovery based mostly on your work. What if the code is based on proprietary or commercial software that can't be shared? And some code is of such miserable quality that scientists find it embarrassing to share.

The Community Research and Academic Programming License (CRAPL), a copyright agreement drafted by Matt Might for use with academic software, includes in its "Definitions" section the following:

2. "The Program" refers to the medley of source code, shell scripts, executables, objects, libraries and build files supplied to You, or these files as modified by You.

    [Any appearance of design in the Program is purely coincidental and should not in any way be mistaken for evidence of thoughtful software construction.]

3. "You" refers to the person or persons brave and daft enough to use the Program.

4. "The Documentation" refers to the Program.

5. "The Author" probably refers to the caffeine-addled graduate student that got the Program to work moments before a submission deadline.

The CRAPL also stipulates that users must "agree to hold the Author free from shame, embarrassment, or ridicule for any hacks, kludges, or leaps of faith found within the Program." While the CRAPL may not be the most legally rigorous licensing agreement, it speaks to the problems faced by authors of academic code: writing software for public use takes a great deal more work than writing code for personal use, including documentation, testing, and cleanup of accumulated cruft from many nights of hacking. The extra work has little benefit for the programmer, who gets no academic credit even for important software that took months to write. And would scientists avail themselves of the opportunity to inspect code and find bugs? Nobody gets scientific glory by checking code for typos.

Experiment, Rinse, Repeat

Another solution might be replication. If scientists carefully re-create the experiments of other scientists from scratch, collecting entirely new data, and validate their results—a painstaking and time-consuming process—it is much easier to rule out the possibility of a typo causing an errant result. Replication also weeds out fluke false positives, assuming the replication attempt has sufficient statistical power to detect the effect in question. Many scientists claim that experimental replication is the heart of science; no new idea is accepted until it is independently tested and retested around the world and found to hold water.

That's not entirely true. Replication is rarely performed for its own sake (except in certain fields—physicists love to make more and more precise measurements of physical constants). Since replicating a complicated result may take months, replication usually happens only when researchers need to use a previous result for their own work. Otherwise, replication is rarely considered publication worthy. Rare exceptions include the Reproducibility Project, born out of increasing concern among psychologists that many important results may not survive replication. Run by a large collaboration of psychologists, the project has been steadily retesting articles from prominent psychology journals. Preliminary results are promising, with most results reproduced in new trials, but there's a long way to go.

In another example, cancer researchers at the pharmaceutical company Amgen retested 53 landmark preclinical studies in cancer research. (By "preclinical" I mean the studies did not involve human patients, because they were testing new and unproven ideas.) Despite working in collaboration with the authors of the original papers, the Amgen researchers could reproduce only six of the studies.[10] Bayer researchers have reported similar difficulties when testing potential new drugs found in published papers.[11]

This is worrisome. Does the trend hold true for less speculative kinds of medical research? Apparently so. Of the top-cited research articles in medicine, a quarter have gone untested after their publication, and a third have been found to be exaggerated or wrong by later research.[12] That's not as extreme as the Amgen result, but it makes you wonder what major errors still lurk unnoticed in important research. Replication is not as prevalent as we would like it to be, and the results are not always favorable.

TIPS
- Automate your data analysis using a spreadsheet, analysis script, or program that can be tested against known input. If anyone suspects an error, you should be able to refer to your code to see exactly what you did.

- Corollary: Test all analysis programs against known input and ensure the results make sense. Ideally, use automated tests to check the code as you make changes, ensuring you don't introduce errors.

- When writing software, follow the best practices for scientific computing: *http://www.plosbiology.org/article/info: doi/10.1371/journal.pbio.1001745*.

- When using programs and scripts to analyze your data, follow the "Ten Simple Rules for Reproducible Computational Research."[9]

- Use a reproducible research tool like Sweave to automatically include data from your analysis in your paper.

- Make all data available when possible, through specialized databases such as GenBank and PDB or through generic data repositories such as Dryad and Figshare.

- Publish your software source code, spreadsheets, or analysis scripts. Many journals let you submit these as supplementary material with your paper, or you can deposit the files on Dryad or Figshare.

11

HIDING THE DATA

I've talked about the common mistakes made by scientists and how the best way to spot them is with a bit of outside scrutiny. Peer reviewers provide some of this scrutiny, but they don't have time to extensively reanalyze data and read code for typos—they can check only that the methodology makes sense. Sometimes they spot obvious errors, but subtle problems are usually missed.[1]

This is one reason why many journals and professional societies require researchers to make their data available to other scientists upon request. Full datasets are usually too large to print in the pages of a journal, and online publication of results is rare—full data is available online for less than 10% of papers published by top journals, with partial publication of select results more common.[2] Instead, authors report their results and send the complete data to other scientists only if they ask for a copy. Perhaps they will find an error or a pattern the original scientists missed, or perhaps they can use the data to investigate a related topic. Or so it goes in theory.

Captive Data

In 2005, Jelte Wicherts and colleagues at the University of Amsterdam decided to analyze every recent article in several prominent journals of the American Psychological Association (APA) to learn about their statistical methods. They chose the APA partly because it requires authors to agree to share their data with other psychologists seeking to verify their claims. But six months later, they had received data for only 64 of the 249 studies they sought it for. Almost three-quarters of authors never sent their data.[3]

Of course, scientists are busy people. Perhaps they simply didn't have time to compile their datasets, produce documents describing what each variable meant and how it was measured, and so on. Or perhaps their motive was self-preservation; perhaps their data was not as conclusive as they claimed. Wicherts and his colleagues decided to test this. They trawled through all the studies, looking for common errors that could be spotted by reading the paper, such as inconsistent statistical results, misuse of statistical tests, and ordinary typos. At least half of the papers had an error, usually minor, but 15% reported at least one statistically significant result that was significant only because of an error.

Next Wicherts and his colleagues looked for a correlation between these errors and an unwillingness to share data. There was a clear relationship. Authors who refused to share their data were more likely to have committed an error in their paper, and their statistical evidence tended to be weaker.[4] Because most authors refused to share their data, Wicherts could not dig for deeper statistical errors, and many more may be lurking.

This is certainly not proof that authors hid their data because they knew their results were flawed or weak; there are many possible confounding factors. Correlation doesn't imply causation, but it does waggle its eyebrows suggestively and gesture furtively while mouthing, "Look over there."* And the surprisingly high error rates demonstrate why data should be

---

*Joke shamelessly stolen from the alternate text of *http://xkcd.com/552/*.

shared. Many errors are not obvious in the published paper and will be noticed only when someone reanalyzes the original data from scratch.

Obstacles to Sharing

Sharing data isn't always as easy as posting a spreadsheet online, though some fields do facilitate it. There are gene sequencing databases, protein structure databanks, astronomical observation databases, and earth observation collections containing the contributions of thousands of scientists. Medical data is particularly tricky, however, since it must be carefully scrubbed of any information that may identify a patient. And pharmaceutical companies raise strong objections to sharing their data on the grounds that it is proprietary. Consider, for example, the European Medicines Agency (EMA).

In 2007, researchers from the Nordic Cochrane Center sought data from the EMA about two weight-loss drugs. They were conducting a systematic review of the effectiveness of the drugs and knew that the EMA, as the authority in charge of allowing drugs onto the European market, would have trial data submitted by the manufacturers that was perhaps not yet published publicly. But the EMA refused to disclose the data on the grounds that it might "unreasonably undermine or prejudice the commercial interests of individuals or companies" by revealing their trial design methods and commercial plans. They rejected the claim that withholding the data could harm patients.

After three and a half years of bureaucratic wrangling and after reviewing each study report and finding no secret commercial information, the European Ombudsman finally ordered the EMA to release the documents. In the meantime, one of the drugs had been taken off the market because of side effects including serious psychiatric problems.[5]

Academics use similar justifications to keep their data private. While they aren't worried about commercial interests, they *are* worried about competing scientists. Sharing a dataset may mean being beaten to your next discovery by a freeloader who acquired the data, which took you months and thousands of dollars to collect, for free. As a result, it is common practice in some fields to consider sharing data only after it is no longer useful to you—once you have published as many papers about it as you can.

Fear of being scooped is a powerful obstacle in academia, where career advancement depends on publishing many papers in prestigious journals. A junior scientist cannot afford to waste six months of work on a project only to be beaten to publication by someone else. Unlike in basketball, there is no academic credit for assists; if you won't get coauthor credit, why bother sharing the data with anyone? While this view is incompatible with the broader goal of the rapid advancement of science, it is compelling for working scientists.

Apart from privacy, commercial concerns, and academic competition, there are practical issues preventing data sharing. Data is frequently stored in unusual formats produced by various scientific instruments or analysis packages, and spreadsheet software saves data in proprietary or incompatible formats. (There is no guarantee that your Excel spreadsheet or SPSS data file will be readable 30 years from now, or even by a colleague using different software.) Not all data can easily be uploaded as a spreadsheet anyway—what about an animal behavior study that recorded hours of video or a psychology study supported by hours of interviews? Even if sufficient storage space were found to archive hundreds of hours of video, who would bear the costs and would anyone bother to watch it?

Releasing data also requires researchers to provide descriptions of the data format and measurement techniques—what equipment settings were used, how calibration was handled, and so on. Laboratory organization is often haphazard, so researchers may not have the time to assemble their collection of spreadsheets and handwritten notes; others may not have a way of sharing gigabytes of raw data.

Data Decay

Another problem is the difficulty of keeping track of data as computers are replaced, technology goes obsolete, scientists move to new institutions, and students graduate and leave labs. If the dataset is no longer in use by its creators, they have no incentive to maintain a carefully organized personal archive of datasets, particularly when data has to be reconstructed from floppy disks and filing cabinets. One study of 516 articles published between 1991 and 2011 found that the probability of data being available decayed over time. For papers more than 20 years old, fewer than half of datasets were available.[6,7] Some authors could not be contacted because their email addresses had changed; others replied that they probably have the data, but it's on a floppy disk and they no longer have a floppy drive or that the data was on a stolen computer or otherwise lost. The decay is illustrated in Figure 11-1.*

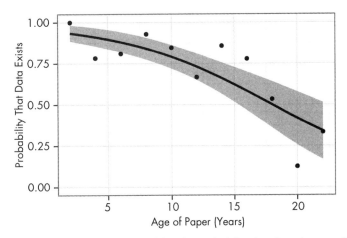

Figure 11-1: As papers get older, the probability that their data is still in existence decays. The solid line is a fitted curve, and the gray band is its 95% confidence band; the points indicate the average availability rates for papers at each age. This plot only includes papers for which authors could be contacted.

Various startups and nonprofits are trying to address this problem. Figshare, for instance, allows researchers to upload gigabytes of data, plots, and presentations to be shared publicly in any file format. To encourage sharing, submissions are given a digital object identifier (DOI), a unique ID commonly used to cite journal articles; this makes it easy to cite the original creators of the data when reusing it, giving them academic credit for their hard work. The Dryad Digital Repository partners with scientific journals to allow authors to deposit data during the article submission process and encourages authors to cite data they have relied on. Dryad promises to convert files to new formats as older formats become obsolete, preventing data from fading into obscurity as programs lose the ability to read it. Dryad also keeps copies of data at several universities to guard against accidental loss.

---

*The figure was produced using code written by the authors of the study, who released it into the public domain and deposited it with their data in the Dryad Digital Repository. Their results will likely last longer than those of the studies they investigated.

The eventual goal is to make it easy to get credit for the publication and reuse of your data. If another scientist uses your data to make an important discovery, you can bask in the reflected glory, and citations of your data can be listed in the same way citations of your papers are. With this incentive, scientists may be able to justify the extra work to deposit their datasets online. But will this be enough? Scientific practice changes very slowly. And will anyone bother to check the data for errors?

Just Leave Out the Details

It is difficult to ask for data you do not know exists. Journal articles are often highly abridged summaries of the years of research they report on, and scientists have a natural prejudice toward reporting the parts that worked. If a measurement or test turned out to be irrelevant to the final conclusions, it will be omitted. If several outcomes were measured and one showed statistically insignificant changes during the study, it won't be mentioned unless the insignificance is particularly interesting.

Journal space limits frequently force the omission of negative results and detailed methodological details. It is not uncommon for major journals to enforce word limits on articles: the *Lancet*, for example, requires articles to be less than 3,000 words, while *Science* limits articles to 4,500 words and suggests that methods be described in an online supplement to the article. Online-only journals such as *PLOS ONE* do not need to pay for printing, so there are no length limits.

Known Unknowns

It's possible to evaluate studies to see what they left out. Scientists leading medical trials are required to provide detailed study plans to ethical review boards before starting a trial, so one group of researchers obtained a collection of these protocols from a Danish review board.[8] The protocols specify how many patients will be recruited, what outcomes will be measured, how missing data (such as patient dropouts or accidental sample losses) will be handled, what statistical analyses will be performed, and so on. Many study protocols had important missing details, however, and few published papers matched the protocols.

We have seen how important it is for studies to collect a sufficiently large sample of data, and most of the ethical review board filings detailed the calculations used to determine an appropriate sample size. However, less than half of the published papers described the sample-size calculation in detail. It also appears that recruiting patients for clinical trials is difficult—half of the studies recruited different numbers of patients than they intended to, and sometimes the researchers did not explain why this happened or what impact it may have on the results.

Worse, many of the scientists omitted results. The review board filings listed outcomes that would be measured by each study: side-effect rates, patient-reported symptoms, and so on. Statistically significant changes in these outcomes were usually reported in published papers, but statistically insignificant results were omitted, as though the researchers had never measured them. Obviously, this leads to hidden multiple comparisons. A study may monitor many outcomes but report only the few that are statistically significant. A casual reader would never know that the study had monitored the insignificant outcomes. When surveyed, most of the researchers denied omitting outcomes, but the review board filings belied their claims. Every paper written by a researcher who denied omitting outcomes had, in fact, left some outcomes unreported.

Outcome Reporting Bias

In medicine, the gold standard of evidence is a meta-analysis of many well-conducted randomized trials. The Cochrane Collaboration, for example, is an international group of volunteers that systematically reviews published randomized trials about various issues in medicine and then produces a report summarizing current knowledge in the field and the treatments and techniques best supported by the evidence. These reports have a reputation for comprehensive detail and methodological scrutiny.

However, if boring results never appear in peer-reviewed publications or are shown in insufficient detail to be of use, the Cochrane researchers will never include them in reviews, causing what is known as *outcome reporting bias*, where systematic reviews become biased toward more extreme and more interesting results. If the Cochrane review is to cover the use of a particular steroid drug to treat pregnant women entering labor prematurely, with the target outcome of interest being infant mortality rates, it's no good if some of the published studies collected mortality data but didn't describe it in any detail because it was statistically insignificant.*

A systematic review of Cochrane systematic reviews revealed that more than a third are probably affected by outcome reporting bias. Reviewers sometimes did not realize outcome reporting bias was present, instead assuming the outcome simply hadn't been measured. It's impossible to exactly quantify how a review's results would change if unpublished results were included, but by their estimate, a fifth of statistically significant review results could become insignificant, and a quarter could have their effect sizes decrease by 20% or more.[9]

Other reviews have found similar problems. Many studies suffer from missing data. Some patients drop out or do not appear for scheduled checkups. While researchers frequently note that data was missing, they frequently do not explain why or describe how patients with incomplete data were handled in the analysis, though missing data can cause biased results (if, for example, those with the worst side effects drop out and aren't counted).[10] Another review of medical trials found that most studies omit important methodological details, such as stopping rules and power calculations, with studies in small specialist journals faring worse than those in large general medicine journals.[11]

Medical journals have begun to combat this problem by coming up with standards, such as the CONSORT checklist, which requires reporting of statistical methods, all measured outcomes, and any changes to the trial design after it began. Authors are required to follow the checklist's requirements before submitting their studies, and editors check to make sure all relevant details are included. The checklist seems to work; studies published in journals that follow the guidelines tend to report more essential detail, though not all of it.[12] Unfortunately, the standards are inconsistently applied, and studies often slip through with missing details.[13] Journal editors will need to make a greater effort to enforce reporting standards.

Of course, underreporting is not unique to medicine. Two-thirds of academic psychologists admit to sometimes omitting some outcome variables in their papers, creating outcome

---

*The Cochrane Collaboration logo is a chart depicting the results of studies on corticosteroids given to women entering labor prematurely. The studies on their own were statistically insignificant, but when the data was aggregated, it was clear the treatment would save lives. This remained undiscovered for years because nobody did a comprehensive review to combine available data.

reporting bias. Psychologists also often report on several experiments in the same paper, testing the same phenomenon from different angles, and half of psychologists admit to reporting only the experiments that worked. These practices persist even though most survey respondents agreed they are probably indefensible.[14]

In biological and biomedical research, the problem often isn't reporting of patient enrollment or power calculations. The problem is the many chemicals, genetically modified organisms, specially bred cell lines, and antibodies used in experiments. Results can be strongly dependent on these factors, but many journals do not have reporting guidelines for these factors, and the majority of chemicals and cells referred to in biomedical papers are not uniquely identifiable, even in journals with strict reporting requirements.[15] Attempt to replicate the findings, as the Bayer and Amgen researchers mentioned earlier did, and you may find it difficult to accurately reproduce the experiment. How can you replicate an immunology paper when it does not state which antibodies to order from the supplier?*

We see that published papers aren't faring very well. What about *unpublished* studies?

Science in a Filing Cabinet

Earlier you saw the impact of multiple comparisons and truth inflation on study results. These problems arise when studies make numerous comparisons with low statistical power, giving a high rate of false positives and inflated estimates of effect sizes, and they appear everywhere in published research.

But not every study is published. We only ever see a fraction of medical research, for instance, because few scientists bother publishing "We Tried This Medicine and It Didn't Seem to Work." In addition, editors of prestigious journals must maintain their reputation for groundbreaking results, and peer reviewers are naturally prejudiced against negative results. When presented with papers with identical methods and writing, reviewers grade versions with negative results more harshly and detect more methodological errors.[16]

---

*I am told that even with the right materials, biological experiments can be fiendishly difficult to reproduce because they are sensitive to tiny variations in experimental setup. But that's not an excuse—it's a serious problem. How can we treat a result as general when it worked only once?

Unpublished Clinical Trials

Consider an example: studies of the tumor suppressor protein TP53 and its effect on head and neck cancer. A number of studies suggested that measurements of TP53 could be used to predict cancer mortality rates since it serves to regulate cell growth and development and hence must function correctly to prevent cancer. When all 18 published studies on TP53 and cancer were analyzed together, the result was a highly statistically significant correlation. TP53 could clearly be measured to tell how likely a tumor is to kill you.

But then suppose we dig up *unpublished* results on TP53: data that had been mentioned in other studies but not published or analyzed. Add this data to the mix, and the statistically significant effect vanishes.[17] After all, few authors bothered to publish data showing no correlation, so the meta-analysis could use only a biased sample.

A similar study looked at reboxetine, an antidepressant sold by Pfizer. Several published studies suggested it was effective compared to a placebo, leading several European countries to approve it for prescription to depressed patients. The German Institute for Quality and Efficiency in Health Care, responsible for assessing medical treatments, managed to get unpublished trial data from Pfizer—three times more data than had ever been published—and carefully analyzed it. The result: reboxetine is not effective. Pfizer had convinced the public that it was effective only by neglecting to mention the studies showing it wasn't.[18]

A similar review of 12 other antidepressants found that of studies submitted to the United States Food and Drug Administration during the approval process, the vast majority of negative results were never published or, less frequently, were published to emphasize secondary outcomes.[19] (For example, if a study measured both depression symptoms and side effects, the insignificant effect on depression might be downplayed in favor of significantly reduced side effects.) While the negative results are available to the FDA to make safety and efficacy determinations, they are not available to clinicians and academics trying to decide how to treat their patients.

This problem is commonly known as *publication bias,* or the *file drawer problem.* Many studies sit in a file drawer for years, never published, despite the valuable data they could contribute. Or, in many cases, studies are published but omit the boring results. If they measured multiple outcomes, such as side effects, they might simply say an effect was "insignificant" without giving any numbers, omit mention of the effect entirely, or quote effect sizes but no error bars, giving no information about the strength of the evidence.

As worrisome as this is, the problem isn't simply the bias on published results. Unpublished results lead to a duplication of effort—if other scientists don't know you've done a study, they may well do it again, wasting money and effort. (I have heard scientists tell stories of talking at a conference about some technique that didn't work, only to find that several scientists in the room had already tried the same thing but not published it.) Funding agencies will begin to wonder why they must support so many studies on the same subject, and more patients and animals will be subjected to experiments.

Spotting Reporting Bias

It is possible to test for publication and outcome reporting bias. If a series of studies have been conducted on a subject and a systematic review has estimated an effect size from the published data, you can easily calculate the power of each individual study in the review.* Suppose, for example, that the effect size is 0.8 (on some arbitrary scale), but the review was composed of many small studies that each had a power of 0.2. You would expect only 20% of the studies to be able to detect the effect—but you may find that 90% or more of the published studies found it because the rest were tossed in the bin.[20]

This test has been used to discover worrisome bias in the publication of neurological studies of animal experimentation.[21] Animal testing is ethically justified on the basis of its benefits to the progress of science and medicine, but evidence of strong outcome reporting bias suggests that many animals have been used in studies that went unpublished, adding nothing to the scientific record.

The same test has been used in a famous controversy in psychology: Daryl Bem's 2011 research claiming evidence for "anomalous retroactive influences on cognition and affect," or psychic prediction of the future. It was published in a reputable journal after peer review but predictably received negative responses from skeptical scientists immediately after publication. Several subsequent papers showed flaws in his analysis

---

*Note that this would not work if each study were truly measuring a different effect because of some systematic differences in how the studies were conducted. Estimating their true power would be much more difficult in that case.

and gave alternative statistical approaches that gave more reasonable results. Some of these are too technically detailed to cover here, but one is directly relevant.

Gregory Francis wondered whether Bem had gotten his good results through publication bias. Knowing his findings would not be readily believed, Bem published not one but 10 different experiments in the same study, with 9 showing statistically significant psychic powers. This would seem to be compelling, but only if there weren't numerous unreported studies that found no psychic powers. Francis found that Bem's success rate did not match his statistical power—it was the result of publication bias, not extrasensory perception.[22]

Francis published a number of similar papers criticizing other prominent studies in psychology, accusing them of obvious publication bias. He apparently trawled through the psychological literature, testing papers until he found evidence of publication bias. This continued until someone noticed the irony.[23] A debate still rages in the psychological literature over the impact of publication bias on publications about publication bias.

Forced Disclosure

Regulators and scientific journals have attempted to halt publication bias. The Food and Drug Administration requires certain kinds of clinical trials to be registered through its website ClinicalTrials.gov before the trials begin, and it requires the publication of summary results on the ClinicalTrials.gov website within a year of the end of the trial. To help enforce registration, the International Committee of Medical Journal Editors announced in 2005 that it would not publish studies that had not been preregistered.

Compliance has been poor. A random sample of all clinical trials registered from June 2008 to June 2009 revealed that more than 40% of protocols were registered *after* the first study participants have been recruited, with the median delinquent study registered 10 months late.[24] This clearly defeats the purpose of requiring advanced registration. Fewer than 40% of protocols clearly specified the primary outcome being measured by the study, the time frame over which it would be measured, and the technique used to measure it, which is unfortunate given that the primary outcome is the *purpose* of the study.

Similarly, reviews of registered clinical trials have found that only about 25% obey the law that requires them to publish results through ClinicalTrials.gov.[25,26] Another quarter of registered trials have no results published *anywhere*, in scientific journals or in the registry.[27] It appears that despite the force of law, most researchers ignore the ClinicalTrials.gov results database and publish in academic journals or not at all; the Food and Drug Admimistration has not fined any drug companies for noncompliance, and journals have not consistently enforced the requirement to register trials.[5] Most peer reviewers do not check trial registers for discrepancies with manuscripts under review, believing this is the responsibility of the journal editors, who do not check either.[28]

Of course, these reporting and registration requirements do not apply to other scientific fields. Researchers in fields such as psychology have suggested encouraging registration by prominently labeling preregistered studies, but such efforts have not taken off.[29] Other suggestions include allowing peer review of study protocols in advance, with the journal deciding to accept or reject the study before data has been collected; acceptance can be based only on the quality of the study's design, rather than its results. But this is not yet widespread. Many studies simply vanish.

TIPS
- Register protocols in public databases, such as ClinicalTrials.gov, the EU Clinical Trials Register (*http://www.clinicaltrialsregister.eu*), or any other public registry. The World Health Organization keeps a list at its International Clinical Trials Registry Platform website (*http://www.who.int/ictrp/en/*), and the SPIRIT checklist (*http://www.spirit-statement.org/*) lists what should be included in a protocol. Post summary results whenever possible.

- Document any deviations from the trial protocol and discuss them in your published paper.

- Make all data available when possible, through specialized databases such as GenBank and PDB or through generic data repositories such as Dryad and Figshare.

- Publish your software source code, Excel workbooks, or analysis scripts used to analyze your data. Many journals will let you submit these as supplementary material with your paper, or you can use Dryad and Figshare.

- Follow reporting guidelines in your field, such as CONSORT for clinical trials, STROBE for observational studies in epidemiology, ARRIVE for animal experiments, or STREGA for gene association studies. The EQUATOR Network (*http://www.equator-network.org/*) maintains lists of guidelines for various fields in medicine.

- If you obtain negative results, publish them! Some journals may reject negative results as uninteresting, so consider open-access electronic-only journals such as *PLOS ONE* or *Trials*, which are peer-reviewed but do not reject studies for being uninteresting. Negative data can also be posted on Figshare.

12

WHAT CAN BE DONE?

I've painted a grim picture. But anyone can pick out small details in published studies and produce a tremendous list of errors. Do these problems matter?

Well, yes. If they didn't, I wouldn't have written this book.

John Ioannidis's famous article "Why Most Published Research Findings Are False"[1] was grounded in mathematical concerns rather than in an empirical test of research results. Since most research articles have poor statistical power and researchers have freedom to choose among analysis methods to get favorable results, while most tested hypotheses are false and most true hypotheses correspond to very small effects, we are mathematically determined to get a plethora of false positives.

But if you want empiricism, you can have it, courtesy of Jonathan Schoenfeld and John Ioannidis. They studied the question "Is everything we eat associated with cancer?"[2,*] After

---

*An important part of the ongoing Oncological Ontology Project to categorize everything into two categories: that which cures cancer and that which causes it.

choosing 50 common ingredients out of a cookbook, they set out to find studies linking them to cancer rates—and found 216 studies on 40 different ingredients. Of course, most of the studies disagreed with each other. Most ingredients had multiple studies alternately claiming they increased and decreased the risk of getting cancer. (Sadly, bacon was one of the few foods consistently found to increase the risk of cancer.) Most of the statistical evidence was weak, and meta-analyses usually showed much smaller effects on cancer rates than the original studies.

Perhaps this is not a serious problem, given that we are already conditioned to ignore news stories about common items causing cancer. Consider, then, a comprehensive review of all research articles published from 2001 to 2010 in the *New England Journal of Medicine*, one of the most prestigious medical research journals. Out of the 363 articles that tested a current standard medical practice, 146 of them—about 40%—concluded that the practice should be abandoned in favor of previous treatments. Only 138 of the studies reaffirmed the current practice.[3]

The astute reader may wonder whether these figures are influenced by publication bias. Perhaps the *New England Journal of Medicine* is biased toward publishing rejections of current standards since they are more exciting. But tests of the current standard of care are genuinely rare and would seem likely to attract an editor's eye. Even if bias does exist, the sheer quantity of these reversals in medical practice should be worrisome.*

Another review compared meta-analyses to subsequent large randomized controlled trials. In more than a third of cases, the randomized trial's outcome did not correspond well to the meta-analysis, indicating that even the careful aggregation of numerous small studies cannot be trusted to give reliable evidence.[4] Other comparisons of meta-analyses found that most results were inflated, with effect sizes decreasing as they were updated with more data. Perhaps a fifth of meta-analysis conclusions represented false positives.[5]

Of course, being contradicted by follow-up studies and meta-analyses doesn't prevent a paper from being used as though it were true. Even effects that have been contradicted

---

*A yet-more-astute reader will ask why we should trust these studies suggesting current practice is wrong, given that so many studies are flawed. That's a fair point, but we are left with massive uncertainty: if we don't know which studies to trust, what *are* the best treatments?

by massive follow-up trials with unequivocal results are frequently cited 5 or 10 years later, with scientists apparently not noticing that the results are false.[6] Of course, new findings get widely publicized in the press, while contradictions and corrections are hardly ever mentioned.[7] You can hardly blame the scientists for not keeping up.

Let's not forget the merely biased results. Poor reporting standards in medical journals mean studies testing new treatments for schizophrenia can neglect to include the scales they used to evaluate symptoms—a handy source of bias because trials using homemade unpublished scales tend to produce better results than those using previously validated tests.[8] Other medical studies simply omit particular results if they're not favorable or interesting, biasing subsequent meta-analyses to include only positive results. A third of meta-analyses are estimated to suffer from this problem.[9]

Multitudes of physical-science papers misuse confidence intervals.[10] And there's a peer-reviewed psychology paper allegedly providing evidence for psychic powers on the basis of uncontrolled multiple comparisons in exploratory studies.[11] Unsurprisingly, the results failed to be replicated—by scientists who appear not to have calculated the statistical power of their tests.[12]

So what can we do? How do we prevent these errors from reaching print? A good starting place would be in statistical education.

Statistical Education

Most American science students have minimal statistical education—one or two required courses at best and none at all for most students. Many of these courses do not cover important concepts such as statistical power and multiple comparisons. And even when students have taken statistics courses, professors report that they can't apply statistical concepts to scientific questions, having never fully understood—or having forgotten—the appropriate techniques. This needs to change. Almost every scientific discipline depends on statistical analysis of experimental data, and statistical errors waste grant funding and researcher time.

It would be tempting to say, "We must introduce a new curriculum adapted to the needs of practicing scientists and require students to take courses in this material" and then assume the problem will be solved. A great deal of research in science education shows that this is not the case. Typical lecture courses teach students little, simply because lectures are a poor way to teach difficult concepts.

Unfortunately, most of this research is not specifically aimed at statistics education. Physicists, however, have done a great deal of research on a similar problem: teaching introductory physics students the basic concepts of force, energy, and kinematics. An instructive example is a large-scale survey of 14 physics courses, including 2,084 students, using the Force Concept Inventory to measure student understanding of basic physics concepts before and after taking the courses. The students began the courses with holes in their knowledge; at the end of the semester, they had filled only 23% of those holes, despite the Force Concept Inventory being regarded as too easy by their instructors.[13]

The results are poor because lectures do not suit how students learn. Students have preconceptions about basic physics from their everyday experience—for example, everyone "knows" that something pushed will eventually come to a stop because every object in the real world does so. But we teach Newton's first law, in which an object in motion stays in motion unless acted upon by an outside force, and expect students to immediately replace their preconception with the new understanding that objects stop only because of frictional forces. Interviews of physics students have revealed numerous surprising misconceptions developed during introductory courses, many not anticipated by instructors.[14,15] Misconceptions are like cockroaches: you have no idea where they came from, but they're everywhere—often where you don't expect them—and they're impervious to nuclear weapons.

We hope that students will learn to solve problems and reason with this new understanding, but usually they don't. Students who watch lectures contradicting their misconceptions report greater confidence in their misconceptions afterward and do no better on simple tests of their knowledge. Often they report not paying attention because the lectures cover concepts they already "know."[16] Similarly, practical demonstrations of physics concepts make little improvement in student understanding because students who misunderstand find ways to interpret the demonstration in light of their misunderstanding.[17] And we can't expect them to ask the right questions in class, because they don't realize they don't understand.

At least one study has confirmed this effect in the teaching of statistical hypothesis testing. Even after reading an article explicitly warning against misinterpreting p values and hypothesis test results in general, only 13% of students correctly answered a questionnaire on hypothesis testing.[18] Obviously, assigning students a book like this one will not be much help if they fundamentally misunderstand statistics. Much of basic statistics is not intuitive (or, at least, not taught in an intuitive fashion), and the opportunity for misunderstanding and error is massive. How can we best teach our students to analyze data and make reasonable statistical inferences?

Again, methods from physics education research provide the answer. If lectures do not force students to confront and correct their misconceptions, we will have to use a method that does. A leading example is peer instruction. Students are assigned readings or videos before class, and class time is spent reviewing the basic concepts and answering conceptual questions. Forced to choose an answer and discuss why they believe it is true *before* the instructor reveals the correct answer, students immediately see when their misconceptions do not match reality, and instructors spot problems before they grow.

Peer instruction has been successfully implemented in many physics courses. Surveys using the Force Concept Inventory found that students typically double or triple their learning gains in a peer instruction course, filling in 50% to 75% of the gaps in their knowledge revealed at the beginning of the semester.[13,19,20] And despite the focus on conceptual understanding, students in peer instruction courses perform just as well or better on quantitative and mathematical questions as their lectured peers.

So far there is relatively little data on the impact of peer instruction in statistics courses. Some universities have experimented with statistics courses integrated with science classes, with students immediately applying statistical knowledge to problems in their field. Preliminary results suggest this works: students learn and retain more statistics, and they spend less time complaining about being forced to take a statistics course.[21] More universities should adopt these techniques and experiment with peer instruction using conceptual tests such as the Comprehensive Assessment of Outcomes in Statistics[22] along with trial courses to see what methods work best. Students will be better prepared for the statistical demands of everyday research if we simply change existing courses, rather than introducing massive new education programs.

But not every student learns statistics in a classroom. I was introduced to statistics when I needed to analyze data in a laboratory and didn't know how; until strong statistics education is more widespread, many students and researchers will find themselves in the same position, and they need resources. The masses of aspiring scientists who Google "how to do a t test" need freely available educational material developed with common errors and applications in mind. Projects like *OpenIntro Statistics*, an open source and freely redistributable introductory statistics textbook, are promising, but we'll need many more. I hope to see more progress in the near future.

Scientific Publishing

Scientific journals are slowly making progress toward solving many of the problems I have discussed. Reporting guidelines, such as CONSORT for randomized trials, make it clear what information is required for a published paper to be reproducible; unfortunately, as you've seen, these guidelines are infrequently enforced. We must continue to pressure journals to hold authors to more rigorous standards.

Premier journals need to lead the charge. *Nature* has begun to do so, announcing a new checklist that authors are required to complete before articles can be published.[23] The checklist requires reporting of sample sizes, statistical power calculations, clinical trial registration numbers, a completed CONSORT checklist, adjustment for multiple comparisons, and sharing of data and source code. The guidelines address most issues covered in this book, except for stopping rules, preferential use of confidence intervals over p values, and discussion of reasons for departing from the trial's registered protocol. *Nature* will also make statisticians available to consult for papers when requested by peer reviewers.

The popular journal *Psychological Science* has recently made similar moves, exempting methods and results sections from article word-count limits and requiring full disclosure of excluded data, insignificant results, and sample-size calculations. Preregistering study protocols and sharing data are strongly encouraged, and the editors have embraced the "new statistics," which emphasizes confidence intervals and effect-size estimates over endless p values.[24] But since confidence intervals are not mandatory, it remains to be seen if their endorsement will make a dent in the established practices of psychologists.

Regardless, more journals should do the same. As these guidelines are accepted by the community, enforcement can follow, and the result will be much more reliable and reproducible research.

There is also much to be said about the unfortunate incentive structures that pressure scientists to rapidly publish small studies with slapdash statistical methods. Promotions, tenure, raises, and job offers are all dependent on having a long list of publications in prestigious journals, so there is a strong incentive to publish promising results as soon as possible. Tenure and hiring committees, composed of overworked academics pushing out their own research papers, cannot extensively review each publication for quality or originality, relying instead on prestige and quantity as approximations. University rankings depend heavily on publication counts and successful grant funding. And because negative or statistically insignificant results will not be published by top journals, it's often not worth the effort to prepare them for publication—publication in lower-class journals may be seen by other academics as a bad sign.

But prestigious journals keep their prestige by rejecting the vast majority of submissions; *Nature* accepts fewer than 10%. Ostensibly this is done because of page limits in the printed editions of journals, though the vast majority of articles are read online. Journal editors attempt to judge which papers will have the greatest impact and interest and consequently choose those with the most surprising, controversial, or novel results. As you've seen, this is a recipe for truth inflation, as well as outcome reporting and publication biases, and strongly discourages replication studies and negative results.

Online-only journals, such as the open-access *PLOS ONE* or BioMed Central's many journals, are not restricted by page counts and have more freedom to publish less obviously exciting articles. But *PLOS ONE* is sometimes seen as a dumping ground for papers that couldn't cut it at more prestigious journals, and some scientists fear publishing in it will worry potential employers. (It's also the single largest academic journal, now publishing more than 30,000 articles annually, so clearly its stigma is not too great.) More prestigious online open-access journals, such as *PLOS Biology* or *BMC Biology*, are also highly selective, encouraging the same kind of statistical lottery.

To spur change, Nobel laureate Randy Schekman announced in 2013 that he and students in his laboratory will no longer publish in "luxury" scientific journals such as *Science* and *Nature*, focusing instead on open-access alternatives (such as *eLife*, which he edits) that do not artificially limit publication by rejecting the vast majority of articles.[25] Of course, Schekman and his students are protected by his Nobel prize, which says more for the quality of his work than the title of the journal it is published in ever could. Average graduate students in average non-Nobel-winning laboratories could not risk damaging their careers with such a radical move.

Perhaps Schekman, shielded by his Nobel, can make the point the rest of us are afraid to make: the frenzied quest for more and more publications, with clear statistical significance and broad applications, harms science. We fixate on statistical significance and do anything to achieve it, even when we don't understand the statistics. We push out numerous small and underpowered studies, padding our résumés, instead of taking the time and money to conduct larger, more definitive ones.

One proposed alternative to the tyranny of prestigious journals is the use of article-level metrics. Instead of judging an article on the prestige of the journal it's published in, judge it on rough measures of its own impact. Online-only journals can easily measure the number of views of an article, the number of citations it has received in other articles, and even how often it is discussed on Twitter or Facebook. This is an improvement over using impact factors, which are a journal-wide average number of citations received by all research articles published in a given year—a self-reinforcing metric since articles from prestigious journals are cited more frequently simply because of their prestige and visibility.

I doubt the solution will be so simple. In open-access journals, article-level metrics reward articles popular among the general public (since open-access articles are free for anyone to read), so an article on the unpleasant composition of chicken nuggets* would score better than an important breakthrough in some arcane branch of genetics. There is no one magic solution; academic culture will have to slowly change to reward the thorough, the rigorous, and the statistically sound.

Your Job

The demands placed on the modern scientist are extreme. Besides mastering their own rapidly advancing fields, most scientists are expected to be good at programming (including

---

*Mostly fat, bone, nerve, and connective tissue, though this article was sadly not actually open-access.[26] The brand of chicken nuggets was not specified.

version control, unit testing, and good software engineering practices), designing statistical graphics, writing scientific papers, managing research groups, mentoring students, managing and archiving data, teaching, applying for grants, and peer-reviewing other scientists' work, along with the statistical skills I'm demanding here. People dedicate their entire careers to mastering one of these skills, yet we expect scientists to be good at all of them to be competitive.

This is nuts. A PhD program can last five to seven years in the United States and still not have time to teach all these skills, except via trial and error.* Tacking on a year or two of experimental design and statistical analysis courses seems unrealistic. Who will have time for it besides statisticians?

Part of the answer is outsourcing. Use the statistical consulting services likely offered by your local statistics department, and rope in a statistician as a collaborator whenever your statistical needs extend beyond a few hours of free advice. (Many statisticians are susceptible to nerd sniping. Describe an interesting problem to them, and they will be unable to resist an attempt at solving it.) In exchange for coauthorship on your paper, the statistician will contribute valuable expertise you can't pick up from two semesters of introductory courses.

Nonetheless, if you're going to do your own data analysis, you'll need a good foundation in statistics, if only to understand what the statistical consultant is telling you. A strong course in applied statistics should cover basic hypothesis testing, regression, statistical power calculation, model selection, and a statistical programming language like R. Or at the least, the course should mention that these concepts exist—perhaps a full mathematical explanation of statistical power won't fit in the curriculum, but students should be aware of power and should know to ask for power calculations when they need them. Sadly, whenever I read the syllabus for an applied statistics course, I notice it fails to cover all of these topics. Many textbooks cover them only briefly.

Beware of false confidence. You may soon develop a smug sense of satisfaction that *your* work doesn't screw up like everyone else's. But I have not given you a thorough introduction to the mathematics of data analysis. There are many ways to foul up statistics beyond these simple conceptual errors. If you're designing an unusual experiment, running a large trial, or analyzing complex data, consult a statistician before you start.

—————————————
*Professional programmers often trade stories about the horrible code produced by self-taught academic friends.

A competent statistician can recommend an experimental design that mitigates issues such as pseudoreplication and helps you collect the right data—and the right quantity of data—to answer your research question. Don't commit the sin, as many do, of showing up to your statistical consultant's office with data in hand, asking, "So how do I tell if this is statistically significant?" A statistician should be a collaborator in your research, not a replacement for Microsoft Excel. You can likely get good advice in exchange for some chocolates or a beer or perhaps coauthorship on your next paper.

Of course, you will do more than analyze your own data. Scientists spend a great deal of time reading papers written by other scientists whose grasp of statistics is entirely unknown. Look for important details in a statistical analysis, such as the following:

- The statistical power of the study or any other means by which the appropriate sample size was determined

- How variables were selected or discarded for analysis

- Whether the statistical results presented support the paper's conclusions

- Effect-size estimates and confidence intervals accompanying significance tests, showing whether the results have practical importance

- Whether appropriate statistical tests were used and, if necessary, how they were corrected for multiple comparisons

- Details of any stopping rules

If you work in a field for which a set of reporting guidelines has been developed (such as the CONSORT checklist for medical trials), familiarize yourself with it and read papers with it in mind. If a paper omits some of the required items, ask yourself what impact that has on its conclusions and whether you can be sure of its results without knowing the missing details. And, of course, pressure journal editors to enforce the guidelines to ensure future papers improve. In fields without standard reporting guidelines, work to create some so that every paper includes all the information needed to evaluate its conclusions.

In short, your task can be expressed in four simple steps.

1. Read a statistics textbook or take a good statistics course. Practice.

2. Plan your data analyses carefully in advance, avoiding the misconceptions and errors I've talked about. Talk to a statistician before you start collecting data.

3. When you find common errors in the scientific literature—such as a simple misinterpretation of p values—hit the perpetrator over the head with your statistics textbook. It's therapeutic.

4. Press for change in scientific education and publishing. It's our research. Let's do it right.

NOTES

Articles from some publishers, such as *BMJ*, *BMC*, and *PLOS*, are freely available online. Free copies of others can sometimes be found by searching for their titles. Most references include Digital Object Identifiers (DOIs), which may be entered at *http://dx.doi.org/* to find the authoritative online version of the article.

Introduction

1. J.P.A. Ioannidis. "Why Most Published Research Findings Are False." *PLOS Medicine* 2, no. 8 (2005): e124. DOI: *10.1371/journal.pmed.0020124.*

2. N.J. Horton and S.S. Switzer. "Statistical Methods in the *Journal*." *New England Journal of Medicine* 353, no. 18 (2005): 1977–1979. DOI: *10.1056/NEJM200511033531823.*

3. B.L. Anderson, S. Williams, and J. Schulkin. "Statistical Literacy of Obstetrics-Gynecology Residents." *Journal of Graduate Medical Education* 5, no. 2 (2013): 272–275. DOI: *10.4300/JGME-D-12-00161.1.*

4. D.M. Windish, S.J. Huot, and M.L. Green. "Medicine residents' understanding of the biostatistics and results in the medical literature." *JAMA* 298, no. 9 (2007): 1010–1022. DOI: *10.1001/jama. 298.9.1010.*

5. S. Goodman. "A Dirty Dozen: Twelve P-Value Misconceptions." *Seminars in Hematology* 45, no. 3 (2008): 135–140. DOI: *10.1053/j. seminhematol.2008.04.003.*

6. P.E. Meehl. "Theory-testing in psychology and physics: A methodological paradox." *Philosophy of Science* 34, no. 2 (1967): 103–115.

7. G. Taubes and C.C. Mann. "Epidemiology faces its limits." *Science* 269, no. 5221 (1995): 164–169. DOI: *10.1126/science.7618077.*

8. D. Fanelli and J.P.A. Ioannidis. "US studies may overestimate effect sizes in softer research." *Proceedings of the National Academy of Sciences* 110, no. 37 (2013): 15031–15036. DOI: *10.1073/pnas. 1302997110.*

Chapter 1

1. B. Thompson. "Two and One-Half Decades of Leadership in Measurement and Evaluation." *Journal of Counseling & Development* 70, no. 3 (1992): 434–438. DOI: *10.1002/j.1556-6676.1992.tb01631.x.*

2. E.J. Wagenmakers. "A practical solution to the pervasive problems of p values." *Psychonomic Bulletin & Review* 14, no. 5 (2007): 779–804. DOI: *10.3758/BF03194105.*

3. J. Neyman and E.S. Pearson. "On the Problem of the Most Efficient Tests of Statistical Hypotheses." *Philosophical Transactions of the Royal Society of London, Series A* 231 (1933): 289–337.

4. S.N. Goodman. "Toward Evidence-Based Medical Statistics. 1: The *P* Value Fallacy." *Annals of Internal Medicine* 130, no. 12 (1999): 995–1004. DOI: *10.7326/0003-4819-130-12-199906150-00008.*

5. S.N. Goodman. "P values, hypothesis tests, and likelihood: implications for epidemiology of a neglected historical debate." *American Journal of Epidemiology* 137, no. 5 (1993): 485–496.

6. R. Hubbard and M.J. Bayarri. "Confusion Over Measures of Evidence (p's) Versus Errors (α's) in Classical Statistical Testing." *The American Statistician* 57, no. 3 (2003): 171–178. DOI: *10.1198/ 0003130031856.*

7. M.J. Gardner and D.G. Altman. "Confidence intervals rather than P values: estimation rather than hypothesis testing." *BMJ* 292 (1986): 746–750.

8. G. Cumming, F. Fidler, M. Leonard, P. Kalinowski, A. Christiansen, A. Kleinig, J. Lo, N. McMenamin, and S. Wilson. "Statistical Reform in Psychology: Is Anything Changing?" *Psychological Science* 18, no. 3 (2007): 230–232. DOI: *10.1111/j.1467-9280.2007.01881.x.*

9. P.E. Tressoldi, D. Giofré, F. Sella, and G. Cumming. "High Impact = High Statistical Standards? Not Necessarily So." *PLOS ONE* 8, no. 2 (2013): e56180. DOI: *10.1371/journal.pone.0056180*.

10. B. Thompson. "Why 'Encouraging' Effect Size Reporting Is Not Working: The Etiology of Researcher Resistance to Changing Practices." *The Journal of Psychology* 133, no. 2 (1999): 133–140. DOI: *10.1080/00223989909599728*.

11. J. Cohen. "The earth is round (p < .05)." *American Psychologist* 49, no. 12 (1994): 997–1003. DOI: *10.1037/0003-066X.49.12.997*.

12. F. Fidler, N. Thomason, G. Cumming, S. Finch, and J. Leeman. "Editors Can Lead Researchers to Confidence Intervals, but Can't Make Them Think: Statistical Reform Lessons From Medicine." *Psychological Science* 15, no. 2 (2004): 119–126. DOI: *10.1111/j.0963-7214.2004.01502008.x*.

Chapter 2

1. P.E. Tressoldi, D. Giofré, F. Sella, and G. Cumming. "High Impact = High Statistical Standards? Not Necessarily So." *PLOS ONE* 8, no. 2 (2013): e56180. DOI: *10.1371/journal.pone.0056180*.

2. R. Tsang, L. Colley, and L.D. Lynd. "Inadequate statistical power to detect clinically significant differences in adverse event rates in randomized controlled trials." *Journal of Clinical Epidemiology* 62, no. 6 (2009): 609–616. DOI: *10.1016/j.jclinepi.2008.08.005*.

3. D. Moher, C. Dulberg, and G. Wells. "Statistical power, sample size, and their reporting in randomized controlled trials." *JAMA* 272, no. 2 (1994): 122–124. DOI: *10.1001/jama.1994.03520020048013*.

4. P.L. Bedard, M.K. Krzyzanowska, M. Pintilie, and I.F. Tannock. "Statistical Power of Negative Randomized Controlled Trials Presented at American Society for Clinical Oncology Annual Meetings." *Journal of Clinical Oncology* 25, no. 23 (2007): 3482–3487. DOI: *10.1200/JCO.2007.11.3670*.

5. C.G. Brown, G.D. Kelen, J.J. Ashton, and H.A. Werman. "The beta error and sample size determination in clinical trials in emergency medicine." *Annals of Emergency Medicine* 16, no. 2 (1987): 183–187. DOI: *10.1016/S0196-0644(87)80013-6*.

6. K.C. Chung, L.K. Kalliainen, and R.A. Hayward. "Type II (beta) errors in the hand literature: the importance of power." *The Journal of Hand Surgery* 23, no. 1 (1998): 20–25. DOI: *10.1016/S0363-5023(98)80083-X*.

7. K.S. Button, J.P.A. Ioannidis, C. Mokrysz, B.A. Nosek, J. Flint, E.S.J. Robinson, and M.R. Munafò. "Power failure: why small sample size undermines the reliability of neuroscience." *Nature Reviews Neuroscience* 14 (2013): 365–376. DOI: *10.1038/nrn3475*.

8. J. Cohen. "The statistical power of abnormal-social psychological research: A review." *Journal of Abnormal and Social Psychology* 65, no. 3 (1962): 145–153. DOI: *10.1037/h0045186*.

9. P. Sedlmeier and G. Gigerenzer. "Do studies of statistical power have an effect on the power of studies?" *Psychological Bulletin* 105, no. 2 (1989): 309–316. DOI: *10.1037/0033-2909.105.2.309*.

10. G. Murray. "The task of a statistical referee." *British Journal of Surgery* 75, no. 7 (1988): 664–667. DOI: *10.1002/bjs.1800750714*.

11. S.E. Maxwell. "The Persistence of Underpowered Studies in Psychological Research: Causes, Consequences, and Remedies." *Psychological Methods* 9, no. 2 (2004): 147–163. DOI: *10.1037/1082-989X.9.2.147*.

12. E. Hauer. "The harm done by tests of significance." *Accident Analysis & Prevention* 36, no. 3 (2004): 495–500. DOI: *10.1016/S0001-4575(03)00036-8*.

13. D.F. Preusser, W.A. Leaf, K.B. DeBartolo, R.D. Blomberg, and M.M. Levy. "The effect of right-turn-on-red on pedestrian and bicyclist accidents." *Journal of Safety Research* 13, no. 2 (1982): 45–55. DOI: *10.1016/0022-4375(82)90001-9*.

14. P.L. Zador. "Right-turn-on-red laws and motor vehicle crashes: A review of the literature." *Accident Analysis & Prevention* 16, no. 4 (1984): 241–245. DOI: *10.1016/0001-4575(84)90019-8*.

15. National Highway Traffic Safety Administration. "The Safety Impact of Right Turn on Red." February 1995. URL: *http://www.nhtsa.gov/people/injury/research/pub/rtor.pdf*.

16. G. Cumming. *Understanding the New Statistics*. Routledge, 2012. ISBN: 978-0415879682.

17. S.E. Maxwell, K. Kelley, and J.R. Rausch. "Sample Size Planning for Statistical Power and Accuracy in Parameter Estimation." *Annual Review of Psychology* 59, no. 1 (2008): 537–563. DOI: *10.1146/annurev.psych.59.103006.093735*.

18. J.P.A. Ioannidis. "Why Most Discovered True Associations Are Inflated." *Epidemiology* 19, no. 5 (2008): 640–648. DOI: *10.1097/EDE.0b013e31818131e7*.

19. J.P.A. Ioannidis. "Contradicted and initially stronger effects in highly cited clinical research." *JAMA* 294, no. 2 (2005): 218–228. DOI: *10.1001/jama.294.2.218*.

20. J.P.A. Ioannidis and T.A. Trikalinos. "Early extreme contradictory estimates may appear in published research: the Proteus phenomenon in molecular genetics research and randomized trials." *Journal of Clinical Epidemiology* 58, no. 6 (2005): 543–549. DOI: *10.1016/j.jclinepi.2004.10.019*.

21. B. Brembs, K.S. Button, and M.R. Munafò. "Deep impact: unintended consequences of journal rank." *Frontiers in Human Neuroscience* 7 (2013): 291. DOI: *10.3389/fnhum.2013.00291*.

22. K.C. Siontis, E. Evangelou, and J.P.A. Ioannidis. "Magnitude of effects in clinical trials published in high-impact general medical journals." *International Journal of Epidemiology* 40, no. 5 (2011): 1280–1291. DOI: *10.1093/ije/dyr095*.

23. A. Gelman and D. Weakliem. "Of beauty, sex, and power: statistical challenges in estimating small effects." *American Scientist* 97 (2009): 310–316. DOI: *10.1511/2009.79.310*.

24. H. Wainer. "The Most Dangerous Equation." *American Scientist* 95 (2007): 249–256. DOI: *10.1511/2007.65.249*.

25. A. Gelman and P.N. Price. "All maps of parameter estimates are misleading." *Statistics in Medicine* 18, no. 23 (1999): 3221–3234. DOI: *10.1002/(SICI)1097-0258(19991215)18:23<3221::AID-SIM312>3.0.CO;2-M*.

26. R. Munroe. "reddit's new comment sorting system." October 15, 2009. URL: *http://redditblog.com/2009/10/reddits-new-comment-sorting-system.html*.

27. E. Miller. "How Not To Sort By Average Rating." February 6, 2009. URL: *http://www.evanmiller.org/how-not-to-sort-by-average-rating.html*.

Chapter 3

1. S.E. Lazic. "The problem of pseudoreplication in neuroscientific studies: is it affecting your analysis?" *BMC Neuroscience* 11 (2010): 5. DOI: *10.1186/1471-2202-11-5*.

2. S.H. Hurlbert. "Pseudoreplication and the design of ecological field experiments." *Ecological Monographs* 54, no. 2 (1984): 187–211. DOI: *10.2307/1942661*.

3. D.E. Kroodsma, B.E. Byers, E. Goodale, S. Johnson, and W.C. Liu. "Pseudoreplication in playback experiments, revisited a decade later." *Animal Behaviour* 61, no. 5 (2001): 1029–1033. DOI: *10.1006/anbe.2000.1676*.

4. D.M. Primo, M.L. Jacobsmeier, and J. Milyo. "Estimating the impact of state policies and institutions with mixed-level data." *State Politics & Policy Quarterly* 7, no. 4 (2007): 446–459. DOI: *10.1177/153244000700700405*.

5. W. Rogers. "Regression standard errors in clustered samples." *Stata Technical Bulletin*, no. 13 (1993): 19–23. URL: *http://www.stata-press.com/journals/stbcontents/stb13.pdf*.

6. L.V. Hedges. "Correcting a Significance Test for Clustering." *Journal of Educational and Behavioral Statistics* 32, no. 2 (2007): 151–179. DOI: *10.3102/1076998606298040*.

7. A. Gelman and J. Hill. *Data Analysis Using Regression and Multilevel/Hierarchical Models.* Cambridge University Press, 2007. ISBN: 978-0521686891.

8.   J.T. Leek, R.B. Scharpf, H.C. Bravo, D. Simcha, B. Langmead, W.E. Johnson, D. Geman, K. Baggerly, and R.A. Irizarry. "Tackling the widespread and critical impact of batch effects in high-throughput data." *Nature Reviews Genetics* 11, no. 10 (2010): 733–739. DOI: *10.1038/nrg2825*.

9.   R.A. Heffner, M.J. Butler, and C.K. Reilly. "Pseudoreplication revisited." *Ecology* 77, no. 8 (1996): 2558–2562. DOI: *10.2307/2265754*.

10.   M.K. McClintock. "Menstrual synchrony and suppression." *Nature* 229 (1971): 244–245. DOI: *10.1038/229244a0*.

11.   H.C. Wilson. "A critical review of menstrual synchrony research." *Psychoneuroendocrinology* 17, no. 6 (1992): 565–591. DOI: *10.1016/0306-4530(92)90016-Z*.

12.   Z. Yang and J.C. Schank. "Women do not synchronize their menstrual cycles." *Human Nature* 17, no. 4 (2006): 433–447. DOI: *10.1007/s12110-006-1005-z*.

13.   A.L. Harris and V.J. Vitzthum. "Darwin's legacy: an evolutionary view of women's reproductive and sexual functioning." *Journal of Sex Research* 50, no. 3-4 (2013): 207–246. DOI: *10.1080/00224499.2012.763085*.

Chapter 4

1.   H. Haller and S. Krauss. "Misinterpretations of significance: A problem students share with their teachers?" *Methods of Psychological Research* 7, no. 1 (2002).

2.   R. Bramwell, H. West, and P. Salmon. "Health professionals' and service users' interpretation of screening test results: experimental study." *BMJ* 333 (2006): 284–286. DOI: *10.1136/bmj.38884.663102.AE*.

3.   D. Hemenway. "Survey Research and Self-Defense Gun Use: An Explanation of Extreme Overestimates." *The Journal of Criminal Law and Criminology* 87, no. 4 (1997): 1430–1445. URL: *http://www.jstor.org/stable/1144020*.

4.   D. McDowall and B. Wiersema. "The incidence of defensive firearm use by US crime victims, 1987 through 1990." *American Journal of Public Health* 84, no. 12 (1994): 1982–1984. DOI: *10.2105/AJPH.84.12.1982*.

5.   G. Kleck and M. Gertz. "Illegitimacy of One-Sided Speculation: Getting the Defensive Gun Use Estimate Down." *Journal of Criminal Law & Criminology* 87, no. 4 (1996): 1446–1461.

6.   E. Gross and O. Vitells. "Trial factors for the look elsewhere effect in high energy physics." *The European Physical Journal C* 70, no. 1-2 (2010): 525–530. DOI: *10.1140/epjc/s10052-010-1470-8*.

7.   E.J. Wagenmakers. "A practical solution to the pervasive problems of p values." *Psychonomic Bulletin & Review* 14, no. 5 (2007): 779–804. DOI: *10.3758/BF03194105*.

8. D.G. Smith, J. Clemens, W. Crede, M. Harvey, and E.J. Gracely. "Impact of multiple comparisons in randomized clinical trials." *The American Journal of Medicine* 83, no. 3 (1987): 545–550. DOI: *10.1016/0002-9343(87)90768-6.*

9. J. Carp. "The secret lives of experiments: methods reporting in the fMRI literature." *Neuroimage* 63, no. 1 (2012): 289–300. DOI: *10.1016/j.neuroimage.2012.07.004.*

10. Y. Benjamini and Y. Hochberg. "Controlling the false discovery rate: a practical and powerful approach to multiple testing." *Journal of the Royal Statistical Society Series B* 57, no. 1 (1995): 289–300. URL: *http://www.jstor.org/stable/2346101.*

Chapter 5

1. A. Gelman and H. Stern. "The Difference Between 'Significant' and 'Not Significant' is not Itself Statistically Significant." *The American Statistician* 60, no. 4 (2006): 328–331. DOI: *10.1198/000313006X152649.*

2. M. Bland. "Keep young and beautiful: evidence for an 'anti-aging' product?" *Significance* 6, no. 4 (2009): 182–183. DOI: *10.1111/j.1740-9713.2009.00395.x.*

3. S. Nieuwenhuis, B.U. Forstmann, and E.J. Wagenmakers. "Erroneous analyses of interactions in neuroscience: a problem of significance." *Nature Neuroscience* 14, no. 9 (2011): 1105–1109. DOI: *10.1038/nn.2886.*

4. A.F. Bogaert. "Biological versus nonbiological older brothers and men's sexual orientation." *Proceedings of the National Academy of Sciences* 103, no. 28 (2006): 10771–10774. DOI: *10.1073/pnas.0511152103.*

5. J. McCormack, B. Vandermeer, and G.M. Allan. "How confidence intervals become confusion intervals." *BMC Medical Research Methodology* 13 (2013). DOI: *10.1186/1471-2288-13-134.*

6. N. Schenker and J.F. Gentleman. "On judging the significance of differences by examining the overlap between confidence intervals." *The American Statistician* 55, no. 3 (2001): 182–186. DOI: *10.1198/000313001317097960.*

7. S. Belia, F. Fidler, J. Williams, and G. Cumming. "Researchers misunderstand confidence intervals and standard error bars." *Psychological methods* 10, no. 4 (2005): 389–396. DOI: *10.1037/1082-989X.10.4.389.*

8. J.R. Lanzante. "A cautionary note on the use of error bars." *Journal of Climate* 18, no. 17 (2005): 3699–3703. DOI: *10.1175/JCLI3499.1.*

9. K.R. Gabriel. "A simple method of multiple comparisons of means." *Journal of the American Statistical Association* 73, no. 364 (1978): 724–729. DOI: *10.1080/01621459.1978.10480084.*

10. M.R. Stoline. "The status of multiple comparisons: simultaneous estimation of all pairwise comparisons in one-way ANOVA designs." *The American Statistician* 35, no. 3 (1981): 134–141. DOI: *10.1080/00031305.1981.10479331.*

Chapter 6

1. P.N. Steinmetz and C. Thorp. "Testing for effects of different stimuli on neuronal firing relative to background activity." *Journal of Neural Engineering* 10, no. 5 (2013): 056019. DOI: *10.1088/1741-2560/10/5/056019.*

2. N. Kriegeskorte, W.K. Simmons, P.S.F. Bellgowan, and C.I. Baker. "Circular analysis in systems neuroscience: the dangers of double dipping." *Nature Neuroscience* 12, no. 5 (2009): 535–540. DOI: *10.1038/nn.2303.*

3. E. Vul, C. Harris, P. Winkielman, and H. Pashler. "Puzzlingly high correlations in fMRI studies of emotion, personality, and social cognition." *Perspectives on Psychological Science* 4, no. 3 (2009): 274–290. DOI: *10.1111/j.1745-6924.2009.01125.x.*

4. E. Vul and H. Pashler. "Voodoo and circularity errors." *Neuroimage* 62, no. 2 (2012): 945–948. DOI: *10.1016/j.neuroimage.2012.01.027.*

5. S.M. Stigler. *Statistics on the Table.* Harvard University Press, 1999. ISBN: 978-0674009790.

6. J.P. Simmons, L.D. Nelson, and U. Simonsohn. "False-Positive Psychology: Undisclosed Flexibility in Data Collection and Analysis Allows Presenting Anything as Significant." *Psychological Science* 22, no. 11 (2011): 1359–1366. DOI: *10.1177/0956797611417632.*

7. D. Bassler, M. Briel, V.M. Montori, M. Lane, P. Glasziou, Q. Zhou, D. Heels-Ansdell, S.D. Walter, and G.H. Guyatt. "Stopping Randomized Trials Early for Benefit and Estimation of Treatment Effects: Systematic Review and Meta-regression Analysis." *JAMA* 303, no. 12 (2010): 1180–1187. DOI: *10.1001/jama.2010.310.*

8. V.M. Montori, P.J. Devereaux, and N. Adhikari. "Randomized trials stopped early for benefit: a systematic review." *JAMA* 294, no. 17 (2005): 2203–2209. DOI: *10.1001/jama.294.17.2203.*

9. S. Todd, A. Whitehead, N. Stallard, and J. Whitehead. "Interim analyses and sequential designs in phase III studies." *British Journal of Clinical Pharmacology* 51, no. 5 (2001): 394–399. DOI: *10.1046/j.1365-2125.2001.01382.x.*

10. L.K. John, G. Loewenstein, and D. Prelec. "Measuring the prevalence of questionable research practices with incentives for truth telling." *Psychological Science* 23, no. 5 (2012): 524–532. DOI: *10.1177/0956797611430953.*

Chapter 7

1. D.G. Altman, B. Lausen, W. Sauerbrei, and M. Schumacher. "Dangers of Using 'Optimal' Cutpoints in the Evaluation of Prognostic Factors." *Journal of the National Cancer Institute* 86, no. 11 (1994): 829–835. DOI: *10.1093/jnci/86.11.829.*

2. L. McShane, D.G. Altman, W. Sauerbrei, S.E. Taube, M. Gion, and G.M. Clark. "Reporting Recommendations for Tumor Marker Prognostic Studies (REMARK)." *Journal of the National Cancer Institute* 97, no. 16 (2005): 1180–1184. DOI: *10.1093/jnci/dji237.*

3. V. Fedorov, F. Mannino, and R. Zhang. "Consequences of dichotomization." *Pharmaceutical Statistics* 8, no. 1 (2009): 50–61. DOI: *10.1002/pst.331.*

4. S.E. Maxwell and H.D. Delaney. "Bivariate Median Splits and Spurious Statistical Significance." *Psychological Bulletin* 113, no. 1 (1993): 181–190. DOI: *10.1037/0033-2909.113.1.181.*

Chapter 8

1. R. Abbaszadeh, A. Rajabipour, M. Mahjoob, M. Delshad, and H. Ahmadi. "Evaluation of watermelons texture using their vibration responses." *Biosystems Engineering* 115, no. 1 (2013): 102–105. DOI: *10.1016/j.biosystemseng.2013.01.001.*

2. M.J. Whittingham, P.A. Stephens, R.B. Bradbury, and R.P. Freckleton. "Why do we still use stepwise modelling in ecology and behaviour?" *Journal of Animal Ecology* 75, no. 5 (2006): 1182–1189. DOI: *10.1111/j.1365-2656.2006.01141.x.*

3. D.A. Freedman. "A note on screening regression equations." *The American Statistician* 37, no. 2 (1983): 152–155. DOI: *10.1080/00031305.1983.10482729.*

4. L.S. Freedman and D. Pee. "Return to a note on screening regression equations." *The American Statistician* 43, no. 4 (1989): 279–282. DOI: *10.1080/00031305.1989.10475675.*

5. R. Investigators and Prevenzione. "Efficacy of n-3 polyunsaturated fatty acids and feasibility of optimizing preventive strategies in patients at high cardiovascular risk: rationale, design and baseline characteristics of the Rischio and Prevenzione study, a large randomised trial in general practice." *Trials* 11, no. 1 (2010): 68. DOI: *10.1186/1745-6215-11-68.*

6. The Risk and Prevention Study Collaborative Group. "n–3 Fatty Acids in Patients with Multiple Cardiovascular Risk Factors." *New England Journal of Medicine* 368, no. 19 (2013): 1800–1808. DOI: *10.1056/NEJMoa1205409.*

7. C. Tuna. "When Combined Data Reveal the Flaw of Averages." *The Wall Street Journal* (2009). URL: *http://online.wsj.com/news/articles/SB125970744553071829.*

8. P.J. Bickel, E.A. Hammel, and J.W. O'Connell. "Sex bias in graduate admissions: Data from Berkeley." *Science* 187, no. 4175 (1975): 398–404. DOI: *10.1126/science.187.4175.398.*

9. S.A. Julious and M.A. Mullee. "Confounding and Simpson's paradox." *BMJ* 309, no. 6967 (1994): 1480–1481. DOI: *10.1136/bmj.309.6967.1480.*

10. R. Perera. "Commentary: Statistics and death from meningococcal disease in children." *BMJ* 332, no. 7553 (2006): 1297–1298. DOI: *10.1136/bmj.332.7553.1297.*

Chapter 9

1. J.P.A. Ioannidis. "Why Most Discovered True Associations Are Inflated." *Epidemiology* 19, no. 5 (2008): 640–648. DOI: *10.1097/EDE.0b013e31818131e7.*

2. M.J. Shun-Shin and D.P. Francis. "Why Even More Clinical Research Studies May Be False: Effect of Asymmetrical Handling of Clinically Unexpected Values." *PLOS ONE* 8, no. 6 (2013): e65323. DOI: *10.1371/journal.pone.0065323.*

3. J.P. Simmons, L.D. Nelson, and U. Simonsohn. "False-Positive Psychology: Undisclosed Flexibility in Data Collection and Analysis Allows Presenting Anything as Significant." *Psychological Science* 22, no. 11 (2011): 1359–1366. DOI: *10.1177/0956797611417632.*

4. A.T. Beall and J.L. Tracy. "Women Are More Likely to Wear Red or Pink at Peak Fertility." *Psychological Science* 24, no. 9 (2013): 1837–1841. DOI: *10.1177/0956797613476045.*

5. A. Gelman. "Too Good to Be True." *Slate* (2013). URL: *http://www.slate.com/articles/health_and_science/science/2013/07/statistics_and_psychology_multiple_comparisons_give_spurious_results.html.*

6. K.M. Durante, A. Rae, and V. Griskevicius. "The Fluctuating Female Vote: Politics, Religion, and the Ovulatory Cycle." *Psychological Science* 24, no. 6 (2013): 1007–1016. DOI: *10.1177/0956797612466416.*

7. C.R. Harris and L. Mickes. "Women Can Keep the Vote: No Evidence That Hormonal Changes During the Menstrual Cycle Impact Political and Religious Beliefs." *Psychological Science* 25, no. 5 (2014): 1147–1149. DOI: *10.1177/0956797613520236.*

8. M. Jeng. "A selected history of expectation bias in physics." *American Journal of Physics* 74 (2006): 578. DOI: *10.1119/1.2186333.*

9. J.R. Klein and A. Roodman. "Blind analysis in nuclear and particle physics." *Annual Review of Nuclear and Particle Science* 55 (2005): 141–163. DOI: *10.1146/annurev.nucl.55.090704.151521.*

10. A.W. Chan, A. Hróbjartsson, K.J. Jørgensen, P.C. Gøtzsche, and D.G. Altman. "Discrepancies in sample size calculations and data analyses reported in randomised trials: comparison of publications with protocols." *BMJ* 337 (2008): a2299. DOI: *10.1136/bmj. a2299.*

11. A.W. Chan, A. Hróbjartsson, M.T. Haahr, P.C. Gøtzsche, and D.G. Altman. "Empirical Evidence for Selective Reporting of Outcomes in Randomized Trials: Comparison of Protocols to Published Articles." *JAMA* 291, no. 20 (2004): 2457–2465. DOI: *10.1001/jama. 291.20.2457.*

12. D. Fanelli and J.P.A. Ioannidis. "US studies may overestimate effect sizes in softer research." *Proceedings of the National Academy of Sciences* 110, no. 37 (2013): 15031–15036. DOI: *10.1073/pnas. 1302997110.*

Chapter 10

1. P.C. Gøtzsche. "Believability of relative risks and odds ratios in abstracts: cross sectional study." *BMJ* 333, no. 7561 (2006): 231–234. DOI: *10.1136/bmj.38895.410451.79.*

2. M. Bakker and J.M. Wicherts. "The (mis)reporting of statistical results in psychology journals." *Behavior Research Methods* 43, no. 3 (2011): 666–678. DOI: *10.3758/s13428-011-0089-5.*

3. E. García-Berthou and C. Alcaraz. "Incongruence between test statistics and P values in medical papers." *BMC Medical Research Methodology* 4, no. 1 (2004): 13. DOI: *10.1186/1471-2288-4-13.*

4. P.C. Gøtzsche. "Methodology and overt and hidden bias in reports of 196 double-blind trials of nonsteroidal antiinflammatory drugs in rheumatoid arthritis." *Controlled Clinical Trials* 10 (1989): 31–56. DOI: *10.1016/0197-2456(89)90017-2.*

5. K.A. Baggerly and K.R. Coombes. "Deriving chemosensitivity from cell lines: Forensic bioinformatics and reproducible research in high-throughput biology." *The Annals of Applied Statistics* 3, no. 4 (2009): 1309–1334. DOI: *10.1214/09-AOAS291.*

6. The Economist. "Misconduct in science: An array of errors." September 2011. URL: *http://www.economist.com/node/21528593.*

7. G. Kolata. "How Bright Promise in Cancer Testing Fell Apart." *New York Times* (2011). URL: *http://www.nytimes.com/2011/07/08/ health/research/08genes.html.*

8. V. Stodden, P. Guo, and Z. Ma. "Toward Reproducible Computational Research: An Empirical Analysis of Data and Code Policy Adoption by Journals." *PLOS ONE* 8, no. 6 (2013): e67111. DOI: *10.1371/journal.pone.0067111.*

9. G.K. Sandve, A. Nekrutenko, J. Taylor, and E. Hovig. "Ten Simple Rules for Reproducible Computational Research." *PLOS Computational Biology* 9, no. 10 (2013): e1003285. DOI: *10.1371/journal. pcbi.1003285.*

10. C.G. Begley and L.M. Ellis. "Drug development: Raise standards for preclinical cancer research." *Nature* 483, no. 7 (2012): 531–533. DOI: *10.1038/483531a*.

11. F. Prinz, T. Schlange, and K. Asadullah. "Believe it or not: how much can we rely on published data on potential drug targets?" *Nature Reviews Drug Discovery* 10 (2011): 328–329. DOI: *10.1038/ nrd3439-c1*.

12. J.P.A. Ioannidis. "Contradicted and initially stronger effects in highly cited clinical research." *JAMA* 294, no. 2 (2005): 218–228. DOI: *10.1001/jama.294.2.218*.

Chapter 11

1. S. Schroter, N. Black, S. Evans, F. Godlee, L. Osorio, and R. Smith. "What errors do peer reviewers detect, and does training improve their ability to detect them?" *Journal of the Royal Society of Medicine* 101, no. 10 (2008): 507–514. DOI: *10.1258/jrsm.2008.080062*.

2. A.A. Alsheikh-Ali, W. Qureshi, M.H. Al-Mallah, and J.P.A. Ioannidis. "Public Availability of Published Research Data in High-Impact Journals." *PLOS ONE* 6, no. 9 (2011): e24357. DOI: *10. 1371/journal.pone.0024357*.

3. J.M. Wicherts, D. Borsboom, J. Kats, and D. Molenaar. "The poor availability of psychological research data for reanalysis." *American Psychologist* 61, no. 7 (2006): 726–728. DOI: *10.1037/0003-066X. 61.7.726*.

4. J.M. Wicherts, M. Bakker, and D. Molenaar. "Willingness to Share Research Data Is Related to the Strength of the Evidence and the Quality of Reporting of Statistical Results." *PLOS ONE* 6, no. 11 (2011): e26828. DOI: *10.1371/journal.pone.0026828*.

5. B. Goldacre. *Bad Pharma: How Drug Companies Mislead Doctors and Harm Patients.* Faber & Faber, 2013. ISBN: 978-0865478008.

6. T.H. Vines, A.Y.K. Albert, R.L. Andrew, F. Débarre, D.G. Bock, M.T. Franklin, K.J. Gilbert, J.S. Moore, S. Renaut, and D.J. Rennison. "The availability of research data declines rapidly with article age." *Current Biology* 24, no. 1 (2014): 94–97. DOI: *10.1016/j.cub. 2013.11.014*.

7. T.H. Vines, A.Y.K. Albert, R.L. Andrew, F. Débarre, D.G. Bock, M.T. Franklin, K.J. Gilbert, J.S. Moore, S. Renaut, and D.J. Rennison. "Data from: The availability of research data declines rapidly with article age." *Dryad Digital Repository* (2013). DOI: *10.5061/ dryad.q3g37*.

8. A.W. Chan, A. Hróbjartsson, M.T. Haahr, P.C. Gøtzsche, and D.G. Altman. "Empirical Evidence for Selective Reporting of Outcomes in Randomized Trials: Comparison of Protocols to Published Articles." *JAMA* 291, no. 20 (2004): 2457–2465. DOI: *10.1001/jama. 291.20.2457*.

9. J.J. Kirkham, K.M. Dwan, D.G. Altman, C. Gamble, S. Dodd, R. Smyth, and P.R. Williamson. "The impact of outcome reporting bias in randomised controlled trials on a cohort of systematic reviews." *BMJ* 340 (2010): c365. DOI: *10.1136/bmj.c365.*

10. W. Bouwmeester, N.P.A. Zuithoff, S. Mallett, M.I. Geerlings, Y. Vergouwe, E.W. Steyerberg, D.G. Altman, and K.G.M. Moons. "Reporting and Methods in Clinical Prediction Research: A Systematic Review." *PLOS Medicine* 9, no. 5 (2012): e1001221. DOI: *10.1371/journal.pmed.1001221.*

11. K. Huwiler-Müntener, P. Jüni, C. Junker, and M. Egger. "Quality of Reporting of Randomized Trials as a Measure of Methodologic Quality." *JAMA* 287, no. 21 (2002): 2801–2804. DOI: *10.1001/jama.287.21.2801.*

12. A.C. Plint, D. Moher, A. Morrison, K. Schulz, D.G. Altman, C. Hill, and I. Gaboury. "Does the CONSORT checklist improve the quality of reports of randomised controlled trials? A systematic review." *Medical Journal of Australia* 185, no. 5 (2006): 263–267.

13. E. Mills, P. Wu, J. Gagnier, D. Heels-Ansdell, and V.M. Montori. "An analysis of general medical and specialist journals that endorse CONSORT found that reporting was not enforced consistently." *Journal of Clinical Epidemiology* 58, no. 7 (2005): 662–667. DOI: *10.1016/j.jclinepi.2005.01.004.*

14. L.K. John, G. Loewenstein, and D. Prelec. "Measuring the prevalence of questionable research practices with incentives for truth telling." *Psychological Science* 23, no. 5 (2012): 524–532. DOI: *10.1177/0956797611430953.*

15. N.A. Vasilevsky, M.H. Brush, H. Paddock, L. Ponting, S.J. Tripathy, G.M. LaRocca, and M.A. Haendel. "On the reproducibility of science: unique identification of research resources in the biomedical literature." *PeerJ* 1 (2013): e148. DOI: *10.7717/peerj.148.*

16. G.B. Emerson, W.J. Warme, F.M. Wolf, J.D. Heckman, R.A. Brand, and S.S. Leopold. "Testing for the presence of positive-outcome bias in peer review: a randomized controlled trial." *Archives of Internal Medicine* 170, no. 21 (2010): 1934–1939. DOI: *10.1001/archinternmed.2010.406.*

17. P.A. Kyzas, K.T. Loizou, and J.P.A. Ioannidis. "Selective Reporting Biases in Cancer Prognostic Factor Studies." *Journal of the National Cancer Institute* 97, no. 14 (2005): 1043–1055. DOI: *10.1093/jnci/dji184.*

18. D. Eyding, M. Lelgemann, U. Grouven, M. Härter, M. Kromp, T. Kaiser, M.F. Kerekes, M. Gerken, and B. Wieseler. "Reboxetine for acute treatment of major depression: systematic review and meta-analysis of published and unpublished placebo and selective serotonin reuptake inhibitor controlled trials." *BMJ* 341 (2010): c4737. DOI: *10.1136/bmj.c4737.*

19. E.H. Turner, A.M. Matthews, E. Linardatos, R.A. Tell, and R. Rosenthal. "Selective publication of antidepressant trials and its influence on apparent efficacy." *New England Journal of Medicine* 358, no. 3 (2008): 252–260. DOI: *10.1056/NEJMsa065779*.

20. J.P.A. Ioannidis and T.A. Trikalinos. "An exploratory test for an excess of significant findings." *Clinical Trials* 4, no. 3 (2007): 245–253. DOI: *10.1177/1740774507079441*.

21. K.K. Tsilidis, O.A. Panagiotou, E.S. Sena, E. Aretouli, E. Evangelou, D.W. Howells, R.A.S. Salman, M.R. Macleod, and J.P.A. Ioannidis. "Evaluation of Excess Significance Bias in Animal Studies of Neurological Diseases." *PLOS Biology* 11, no. 7 (2013): e1001609. DOI: *10.1371/journal.pbio.1001609*.

22. G. Francis. "Too good to be true: Publication bias in two prominent studies from experimental psychology." *Psychonomic Bulletin & Review* 19, no. 2 (2012): 151–156. DOI: *10.3758/s13423-012-0227-9*.

23. U. Simonsohn. "It Does Not Follow: Evaluating the One-Off Publication Bias Critiques by Francis." *Perspectives on Psychological Science* 7, no. 6 (2012): 597–599. DOI: *10.1177/1745691612463399*.

24. R.F. Viergever and D. Ghersi. "The Quality of Registration of Clinical Trials." *PLOS ONE* 6, no. 2 (2011): e14701. DOI: *10.1371/journal.pone.0014701*.

25. A.P. Prayle, M.N. Hurley, and A.R. Smyth. "Compliance with mandatory reporting of clinical trial results on ClinicalTrials.gov: cross sectional study." *BMJ* 344 (2012): d7373. DOI: *10.1136/bmj.d7373*.

26. V. Huser and J.J. Cimino. "Linking ClinicalTrials.gov and PubMed to Track Results of Interventional Human Clinical Trials." *PLOS ONE* 8, no. 7 (2013): e68409. DOI: *10.1371/journal.pone.0068409*.

27. C.W. Jones, L. Handler, K.E. Crowell, L.G. Keil, M.A. Weaver, and T.F. Platts-Mills. "Non-publication of large randomized clinical trials: cross sectional analysis." *BMJ* 347 (2013): f6104. DOI: *10.1136/bmj.f6104*.

28. S. Mathieu, A.W. Chan, and P. Ravaud. "Use of trial register information during the peer review process." *PLOS ONE* 8, no. 4 (2013): e59910. DOI: *10.1371/journal.pone.0059910*.

29. E.J. Wagenmakers, R. Wetzels, D. Borsboom, H.L.J. van der Maas, and R.A. Kievit. "An Agenda for Purely Confirmatory Research." *Perspectives on Psychological Science* 7, no. 6 (2012): 632–638. DOI: *10.1177/1745691612463078*.

Chapter 12

1. J.P.A. Ioannidis. "Why Most Published Research Findings Are False." *PLOS Medicine* 2, no. 8 (2005): e124. DOI: *10.1371/journal.pmed.0020124*.

2. J.D. Schoenfeld and J.P.A. Ioannidis. "Is everything we eat associated with cancer? A systematic cookbook review." *American Journal of Clinical Nutrition* 97, no. 1 (2013): 127–134. DOI: *10.3945/ajcn.112.047142.*

3. V. Prasad, A. Vandross, C. Toomey, M. Cheung, J. Rho, S. Quinn, S.J. Chacko, D. Borkar, V. Gall, S. Selvaraj, N. Ho, and A. Cifu. "A Decade of Reversal: An Analysis of 146 Contradicted Medical Practices." *Mayo Clinic Proceedings* 88, no. 8 (2013): 790–798. DOI: *10.1016/j.mayocp.2013.05.012.*

4. J. LeLorier, G. Gregoire, and A. Benhaddad. "Discrepancies between meta-analyses and subsequent large randomized, controlled trials." *New England Journal of Medicine* 337 (1997): 536–542. DOI: *10.1056/NEJM199708213370806.*

5. T.V. Pereira and J.P.A. Ioannidis. "Statistically significant meta-analyses of clinical trials have modest credibility and inflated effects." *Journal of Clinical Epidemiology* 64, no. 10 (2011): 1060–1069. DOI: *10.1016/j.jclinepi.2010.12.012.*

6. A. Tatsioni, N.G. Bonitsis, and J.P.A. Ioannidis. "Persistence of Contradicted Claims in the Literature." *JAMA* 298, no. 21 (2007): 2517–2526. DOI: *10.1001/jama.298.21.2517.*

7. F. Gonon, J.P. Konsman, D. Cohen, and T. Boraud. "Why Most Biomedical Findings Echoed by Newspapers Turn Out to be False: The Case of Attention Deficit Hyperactivity Disorder." *PLOS ONE* 7, no. 9 (2012): e44275. DOI: *10.1371/journal.pone.0044275.*

8. M. Marshall, A. Lockwood, C. Bradley, C. Adams, C. Joy, and M. Fenton. "Unpublished rating scales: a major source of bias in randomised controlled trials of treatments for schizophrenia." *The British Journal of Psychiatry* 176, no. 3 (2000): 249–252. DOI: *10.1192/bjp.176.3.249.*

9. J.J. Kirkham, K.M. Dwan, D.G. Altman, C. Gamble, S. Dodd, R. Smyth, and P.R. Williamson. "The impact of outcome reporting bias in randomised controlled trials on a cohort of systematic reviews." *BMJ* 340 (2010): c365. DOI: *10.1136/bmj.c365.*

10. J.R. Lanzante. "A cautionary note on the use of error bars." *Journal of Climate* 18, no. 17 (2005): 3699–3703. DOI: *10.1175/JCLI3499.1.*

11. E. Wagenmakers, R. Wetzels, D. Borsboom, and H.L. van der Maas. "Why psychologists must change the way they analyze their data: The case of psi." *Journal of Personality and Social Psychology* 100, no. 3 (2011): 426–432. DOI: *10.1037/a0022790.*

12. J. Galak, R.A. LeBoeuf, L.D. Nelson, and J.P. Simmons. "Correcting the past: Failures to replicate psi." *Journal of Personality and Social Psychology* 103, no. 6 (2012): 933–948. DOI: *10.1037/a0029709.*

13. R. Hake. "Interactive-engagement versus traditional methods: A six-thousand-student survey of mechanics test data for introductory physics courses." *American Journal of Physics* 66, no. 1 (1998): 64–74. DOI: *10.1119/1.18809*.

14. L.C. McDermott. "Research on conceptual understanding in mechanics." *Physics Today* 37, no. 7 (1984): 24. DOI: *10.1063/1.2916318*.

15. J. Clement. "Students' preconceptions in introductory mechanics." *American Journal of Physics* 50, no. 1 (1982): 66–71. DOI: *10.1119/1.12989*.

16. D.A. Muller. *Designing Effective Multimedia for Physics Education.* PhD thesis. University of Sydney, April 2008. URL: *http://www.physics.usyd.edu.au/super/theses/PhD(Muller).pdf*.

17. C.H. Crouch, A.P. Fagen, J.P. Callan, and E. Mazur. "Classroom demonstrations: Learning tools or entertainment?" *American Journal of Physics* 72, no. 6 (2004): 835–838. DOI: *10.1119/1.1707018*.

18. H. Haller and S. Krauss. "Misinterpretations of significance: A problem students share with their teachers?" *Methods of Psychological Research* 7, no. 1 (2002).

19. C.H. Crouch and E. Mazur. "Peer Instruction: Ten years of experience and results." *American Journal of Physics* 69, no. 9 (2001): 970–977. DOI: *10.1119/1.1374249*.

20. N. Lasry, E. Mazur, and J. Watkins. "Peer instruction: From Harvard to the two-year college." *American Journal of Physics* 76, no. 11 (2008): 1066–1069. DOI: *10.1119/1.2978182*.

21. A.M. Metz. "Teaching Statistics in Biology: Using Inquiry-based Learning to Strengthen Understanding of Statistical Analysis in Biology Laboratory Courses." *CBE Life Sciences Education* 7 (2008): 317–326. DOI: *10.1187/cbe.07--07--0046*.

22. R. Delmas, J. Garfield, A. Ooms, and B. Chance. "Assessing students' conceptual understanding after a first course in statistics." *Statistics Education Research Journal* 6, no. 2 (2007): 28–58.

23. Nature Editors. "Reporting checklist for life sciences articles." May 2013. URL: *http://www.nature.com/authors/policies/checklist.pdf*.

24. E. Eich. "Business Not as Usual." *Psychological Science* 25, no. 1 (2014): 3–6. DOI: *10.1177/0956797613512465*.

25. R. Schekman. "How journals like Nature, Cell and Science are damaging science." *The Guardian* (2013). URL: *http://www.theguardian.com/commentisfree/2013/dec/09/how-journals-nature-science-cell-damage-science*.

26. R.D. deShazo, S. Bigler, and L.B. Skipworth. "The Autopsy of Chicken Nuggets Reads 'Chicken Little'." *American Journal of Medicine* 126, no. 11 (2013): 1018–1019. DOI: *10.1016/j.amjmed.2013.05.005*.

INDEX

Symbols

α (false positive rate), 11–12

A

accuracy in parameter
        estimation (AIPE), 23
Akaike information criterion, 83
alternative hypothesis, 11
*American Journal of Public
        Health*, 14
American Psychological
        Association (APA), 106
Amgen, 102–103, 113
*An Introduction to Error
        Analysis*, 61
analysis of variance
        (ANOVA), 77
animal testing, 115
antidepressants, 114
ARRIVE guidelines, 118
article-level metrics, 126
assurance, 23
autocorrelation, 33

B

backward elimination, 82–83
Bad Pharma, 4
Baggerly, Keith, 98–99
base rate fallacy, 39–47
    and cancer medication,
        39–40
    and gun use, 45–47
    and mammograms, 42–43
    and smoking, 43–45
Bayer, 103, 113
Bayesian information
        criterion, 83
Bem, Daryl, 115–116

Benjamini–Hochberg
        procedure, 52–53
bias, 91–95
    avoiding, 93–95
    outcome reporting, 111–113
    publication, 114-117
    unconscious, 93
biased coin, 15–18
BioMed Central, 125
bird calls, 32, 33
blind analysis, 93–94
blood pressure, 32–33, 67
body mass index, 75–76
Bonferroni correction, 50,
        51–52
breast cancer, 75
British Journal of Dermatology, 57
Brownlee, K.A., 44–45

C

cancer, 19, 98–99, 102–103, 114
    and base rate fallacy, 39–40
    breast cancer, 42–45, 75
    and food, association with,
        119–120
    and gene expression, 34–35
    kidney cancer, 27–28
causation, and correlation,
        84–85
chicken nuggets, 126
cholesterol, 58, 84
circular analysis, 64–66
climate science, 61
clinical trial protocol, 94
    databases, 117–118
    and preventing false
        positives, 70–71
    registration, 116–117
    reporting, 110–111

ClinicalTrials.gov, 116–117
clustered standard errors, 34
Cochrane Collaboration,
        111–112
Cohen, Jacob, 20, 28
cold medicines, 8–9, 12–13
Community Research and
        Academic Programming
        License (CRAPL),
        101–102
Comprehensive Assessment of
        Outcomes in
        Statistics, 123
confidence interval, 12–14
    mandatory reporting, 14
    and new statistics, 124
    overlap of, 59–62
    overprecise figures, 44
    and precision, 22–23
    pseudoreplication, adjusting
        for, 34
    and reddit voting, 28
    upper bounds, calculating, 23
confounding variable, 76–78,
        80, 84
    and health-care quality, 77
    Simpson's paradox, 85–88
CONSORT checklist, 112, 118,
        124, 128
Continental Airlines, 87–88
Coombes, Kevin, 98–99
correlation, and causation,
        84–85
CRAPL (Community Research
        and Academic
        Programming License),
        101–102
cross-validation, 83

D

descision making in statistical
        analysis, 89–95
dichotomization, 74–78
    and breast cancer, 75
    and obesity, 75–76
difference in significance, 55–58

digital object identifier
        (DOI), 109
double-dipping, 64–71
    mitigation, 66
Dryad Digital Repository, 103,
        109–110, 117
Duke University, 98–99
Dunnington, Frank, 93

E

effect size, 9. *See also* confidence
        interval
    effect on power, 17
    and gender ratios, 25
    and new statistics, 124
    shrinkage, 27–28
electrodes, 64–65
electron charge, 93
eLife, 126
EMA (European Medicines
        Agency), 107
Epidemiology, 14
EQUATOR Network, 118
error bars, 59–62. *See also*
        confidence interval
European Ombudsman, 107
exploratory analysis, 63

F

false discovery rate, 40
    controlling, 52–53
false negative rate, 11–12, 46
false positive rate, 11–12, 46
    and multiple comparisons,
        47–51, 92
    stepwise regression, effect
        on, 82
    stopping rules, effect on,
        68–70
Figshare, 103, 109, 117
file drawer problem. *See*
        publication bias
fish oil, 84–85
Fisher, R.A., 11–12
Fixitol and Solvix example, 18,
        23–24, 56, 69–70

flight delays, 87–88
fMRI, 51–52
   of Atlantic salmon, 51
   and double-dipping, 65–66
Food and Drug Administration
   (FDA), 114, 116
Force Concept Inventory,
   122, 123
forward selection, 82
Francis, Gregory, 116

G

Gabriel comparison intervals,
   61–62
Galton, Francis, 67
Gelman, Andrew, 93
GenBank, 103, 117
gender discrimination, in
   graduate admissions,
   85–86
gender ratios, 25
gene association studies, 24, 118
gene expression, 34–35
genetics, 24, 98–99
Goldacre, Ben, 4
graduate admissions, gender
   discrimination in, 85–86
Graham, Paul, xvi
gun control, 45–47

H

Hanlon's razor, 4
health-care quality, 77
heart attack, 20, 84–85
hierarchical models, 34
Higgs boson, 41, 49–50
homosexuality, 58
Hotelling, Harold, 68
How to Lie with Smoking Statistics,
   43–45
How to Lie with Statistics, 1, 43
Huff, Darrell, 1, 43–45
hypothesis
   alternative, 11
   null, 11–12
hypothesis testing. *See p* value

I

impact factor, 25, 126
Institute for Quality and
   Efficiency in Health
   Care, 114
International Committee of
   Medical Journal
   Editors, 116
Ioannidis, John, 119
IPython Notebook, 100
IQ test, 18

J

jelly beans, 47–49
Journal of Abnormal and Social
   Psychology, 20
Journal of the American Statistical
   Association, 68
Journal of Theoretical Biology, 26

K

Kanazawa, Satoshi, 25–26
kidney cancer, 27–28
kidney stones, 86–87

L

Lancet, 110
Large Hadron Collider, 41, 49
lasso (least absolute shrinkage
   and selection
   operator), 84
LaTeX, 100
leave-out-one cross-
   validation, 83
look-elsewhere effect, 49–50

M

mammograms, 42–43
McClintock, Martha, 36
measurement error, 17–18
median split, 74
mediocrity. *See The Triumph of*
   Mediocrity in Business
Meehl, Paul, 3

meningitis, 87
menstrual cycles, 92–93
    synchronization of, 35–38
meta-analyses, 120
    and outcome reporting bias,
        111–112
    on statin drug research, 58
microarrays, 34–35, 98
middlebrow dismissals, xvi–xvii
Might, Matt, 101
missing data, 90, 112
mistakes, 97–98
multiple comparisons, 47–51
    of Atlantic salmon fMRI,
        51–52
    and circular analysis, 66
    and false discovery rate,
        52–53
    in stepwise regression, 82
    and stopping rules, 69–70

N

National Cancer Institute, 99
Nature, 13, 19, 25, 97, 124, 126
NCVS (National Crime
    Victimization Survey),
    45–47
negative binomial
    distribution, 10
New England Journal of Medicine,
    2, 120
new statistics, 124
Neyman, Jerzy, 11–12
Neyman-Pearson framework,
    11–12
No. 7 Protect & Perfect Beauty
    Serum, 57–58
Nordic Cochrane Center, 107
null hypothesis, 11–12

O

obesity, 73–76
omega-3 fatty acids, 84
Oncological Ontology
    Project, 119
open data, 105–113

OpenIntro Statistics, 124
outcome reporting bias,
    111–113
    tests for, 115–116
overfitting, 82
overprecise figures, 44
ovulation, 92–93. *See also*
    menstrual cycles

P

p value, 8–10
    and base rate fallacy, 40–43
    Bonferroni correction,
        calculating with, 50
    dichotomization, effect on,
        74–75
    and difference in
        significance, 55–58
    double-dipping, effect on, 65
    errors in calculation, 97, 106
    multiple comparisons. *See*
        multiple comparisons
    overuse of, 13
    pseudoreplication, effect on,
        32–34
    and psychic statistics, 10
    quiz, 41–42
    stopping rules, effect on,
        69–70
    versus confidence intervals,
        12–14
PDB, 103, 117
Pearson, Egon, 11–12
peer instruction, 123
penguins, 65–66
penicillin, 87
percutaneous nephroli-
    thotomy, 86
periods. *See* menstural cycles
Pfizer, 114
physics education, 122–123
PLOS ONE, 110, 118, 125
Potti, Anil, 99
power, 15–18
    Bonferroni correction, effect
        on, 50

dichotomization, effect on, 75–76
and differences in significance, 56
power curve, 16–18
underpowered studies, 18–21
practical significance, 9, 13
principal components analysis, 35
protocol. *See* clinical trial protocol
pseudoreplication, 31–38
and bird calls, 33
and menstrual cycle synchronization, 35–38
and microarray processing, 34–35
psychic *p* values, 10
psychic powers, 115–116, 121
Psychological Science, 124
publication bias, 114–117
and antidepressants, 114
avoiding, 116–117
and psychic powers, 115–116
Python programming language, 100

R

R programming language, 100
random assignment, 31, 33, 87
randomized controlled trial, 31
reboxetine, 114
reddit voting, 28
registered studies, 70–71, 116–117
regression modeling, 74, 79–85
evaluating fairly, 83
and heart attacks, 84–85
and stepwise regression, 82–84
and test scores, 80
and watermelon ripeness, 80–82
regression to the mean, 67–68
repeated measures, 34

replication studies, 57, 95, 102–103
reporting guidelines, 112, 124
Reproducibility Project, 102
reproducible research, 99–103
researcher freedom, 89–95
right turn on red, 21–22
Rothman, Kenneth, 14

S

S-phase fraction, 75
salmon, 51–52, 65
sample size, 8
and confidence interval, 23
effect on power, 17–18
and high variance, 26–28
and pseudoreplication, 32–33
and truth inflation, 23
Schekman, Randy, 125–126
Schoenfeld, Jonathan, 119
Science, 19, 25, 110, 125
Secrist, Horace, 68
sequential analysis, 70
shrinkage, 27–28
significance testing. *See p* value
Simpson's paradox, 85–88
and flight delays, 87–88
and gender discrimination, 85–86
and kidney stones, 86–87
and meningitis, 87
Smoking and Health, 43
software, statistical, 100–101
Solvix and Fixitol example, 18, 23–24, 56, 69–70
speed of light, 93
sphygmomanometers, 32
spontaneous human combustion, 13
standard deviation, 60–62
standard error, 60–62
statin drugs, 58
statistical education, 2, 121–124
outside classroom, 123
peer instruction, 123
statistical power. *See* power

Statistical Power Analysis for the Behavioral Sciences, 28
statistical software, 100–101
statistically significant. *See p* value
stepwise regression, 81–84
Stigler's law of eponymy, 85
stopping rules, 68–71
  in truth inflation, 70
  unreported, 112
STREGA guidelines, 118
STROBE guidelines, 118
Surgeon General, 43
Sweave, 100

T

test scores, 26, 68
Thompson, Bruce, 9
TP53 suppressor protein, 114
traffic safety, 21–22
Trials, 118
triglycerides, 84
triple blinding, 94
The Triumph of Mediocrity in Business, 68
Trivers–Willard Hypothesis, 25
truth inflation, 23–26, 63
  and double-dipping, 66
  in model selection, 82
  in replication studies, 57
  stopping rules, 70
turn signals, 37
type M error. *See* truth inflation

U

unconscious bias, 93
underpowered studies, 18–21
United Airlines, 87–88
United States Preventive Services Task Force, 42
University of California, Berkeley, 85–86

V

visual cortex, 64
von Moltke, Helmuth, 95
voxels, 51

W

walruses, 65–66
watermelon ripeness, 80–84
weight-loss drugs, 107
Wicherts, Jelte, 106
winner's curse. *See* truth inflation
wrinkle cream, 57–58

Y

yachts, 77
yellow-bellied sapsucker, 33

The fonts used in *Statistics Done Wrong* are New Baskerville, Futura, TheSansMono Condensed, and Dogma. The book was typeset with LaTeX 2_ε package nostarch by Boris Veytsman *(2008/06/06 v1.3 Typesetting books for No Starch Press).*

Plots were produced with the R statistical programming language (version 3.0.1 "Good Sport"). No statisticians were harmed in the making of this book, though several guests and friends were seriously bored.

The book was printed and bound by Sheridan Books, Inc. in Chelsea, Michigan. The paper is 70# Finch Offset, which is certified by the Forest Stewardship Council (FSC).

UPDATES

Visit *http://www.nostarch.com/statsdonewrong/* for updates, errata, and other information.